The Greeks

Roy Burrell

Illustrated by Peter Connolly

Oxford University Press

Oxford University Press, Great Clarendon Street, Oxford OX2 6DP

Oxford New York
Athens Auckland Bangkok Bogota Bombay
Buenos Aires Calcutta Cape Town Dar es Salaam Delhi
Florence Hong Kong Istanbul Karachi
Kuala Lumpur Madras Madrid Melbourne
Mexico City Nairobi Paris Singapore
Taipei Tokyo Toronto

and associated companies in
Berlin Ibadan

Oxford is a trade mark of Oxford University Press

First published in the UK in paperback and hardback 1989

Reprinted in paperback 1991, 1992, 1997
Reprinted in hardback 1992, 1996, 1997

Library of Congress Catalog Card Nubmber 89–43665

The CIP catalogue record for this book is available from
the British Library
  ISBN 0–19–917161–0 hbk
  ISBN 0–19–917101–7 pbk

Typeset by MS Filmsetting Limited, Frome, Somerset
Printed in Hong Kong

Acknowledgements

The illustrations including the cover and map models are by
Peter Connolly, except for p. 59 (top) which is by John
Batchelor. Handwriting is by Elitta Fell.

The publishers would like to thank the following for permission to
reproduce photographs:

American School of Classical Studies at Athens p. 38 right;
Archaeological Museum of Thessalonika p. 93 left & right;
Ashmolean Museum, Oxford p. 8, p. 8/9, p. 11; John Boardman
p. 53; Mensun Bound, Maritime Archaeological Research,
Oxford p. 47; The Trustees of the British Museum p. 44 left &
right, p. 45 bottom right & top left, p. 60, p. 67, p. 88 top left,
p. 99, p. 111; Nobby Clark p. 73; Peter Connolly p. 27, p. 29;
C M Dixon Photo Resources p. 13 right; Fitzwilliam Museum,
Cambridge p. 20; Sonia Halliday Photographs p. 10, p. 12;
Robert Harding Picture Library p. 37; Hirmer Fotoarchiv p. 13
left, p. 17 top, p. 18, p. 21 top & bottom, p. 41, p. 48 middle &
right; Michael Holford Photographs p. 38 middle, p. 74 left,
p. 78 top, p. 88 top right; Malta High Commission p. 74 right;
Mansell Collection p. 26, p. 48 left; Musées Nationaux p. 83
top; Museum of Fine Arts, Boston p. 45 top right & bottom left;
National Archaeological Museum of Athens p. 17 bottom,
p. 61, p. 82; Ronald Sheridan's Photo-Library p. 38 left, p. 78
bottom, p. 79, p. 88 bottom right & bottom left, p. 110 left &
right.

CONTENTS

The legend of the Minotaur

ONCE UPON A TIME, KING MINOS RULED OVER CRETE. RULERS OF NEARBY LANDS WERE AFRAID OF HIM.

GREECE

Crete

IN AN IMMENSE MAZE UNDER HIS PALACE, HE KEPT A MONSTER CALLED THE MINOTAUR – HALF MAN AND HALF BULL.

EVERY FEW YEARS, THE MINOTAUR WAS FED ON SEVEN YOUNG MEN AND SEVEN YOUNG WOMEN FROM ATHENS. IF THEY WERE NOT SENT, MINOS WOULD HAVE INVADED THEIR LAND.

ONE AT A TIME, THE YOUNG PEOPLE WERE FORCED INTO THE DARKNESS OF THE UNDERGROUND MAZE WHERE THE BULL-MAN COULD BE HEARD ROARING.

YOUNG THESEUS PLEADED WITH HIS FATHER, AEGEUS, KING OF ATHENS, TO LET HIM TAKE THE PLACE OF ONE OF THE VICTIMS.

THE KING AGREED, BUT WITH A HEAVY HEART. 'IF YOU SURVIVE,' HE SAID, 'CHANGE THE SHIP'S BLACK SAILS FOR WHITE ONES WHEN YOU COME HOME. THEN I'LL KNOW WHAT'S HAPPENED EVEN WHILE YOUR VESSEL IS A GREAT WAY OFF!'

ARIADNE, THE DAUGHTER OF MINOS, FELL IN LOVE WITH THESEUS AND SECRETLY GAVE HIM A SWORD WITH WHICH TO KILL THE MINOTAUR AND ALSO A BALL OF WOOL SO THAT HE WOULDN'T LOSE HIS WAY IN THE MAZE.

5

AS SOON AS THESEUS WAS INSIDE THE MAZE, HE TIED ONE END OF THE WOOL TO A SPUR OF ROCK AND TOOK THE SWORD FROM WHERE IT WAS HIDDEN BENEATH HIS TUNIC.

THESEUS FOUND THE MINOTAUR BY ITS ROARING AND ATTACKED IT. THE MONSTER WAS SURPRISED AND ALARMED TO FIND THAT ITS NEXT 'MEAL' WAS ARMED. AFTER A TREMENDOUS FIGHT, THESEUS MANAGED TO KILL THE BEAST.

AAARGH!

THESEUS REWOUND THE WOOL AND THUS FOUND HIS WAY BACK TO THE OUTSIDE WORLD.

AS THE LOVERS FLED TOWARDS THE HARBOUR, IT SEEMED THAT THEY COULD STILL HEAR THE MONSTER ROARING. IN FACT, IT WAS AN EARTHQUAKE THAT SMASHED THE CRETAN KING'S PALACE AT KNOSSOS. THE FIRES THAT BURNT IT TO THE GROUND SIGNALLED THE END OF CRETE'S POWER.

THESEUS WAS SO DELIGHTED AT OVERCOMING THE MONSTER AND SO MUCH IN LOVE WITH THE PRINCESS THAT HE FORGOT ALL ABOUT CHANGING THE BLACK SAIL FOR A WHITE ONE.

AEGEUS, THE FATHER OF THESEUS, SAW THE SHIP IN THE DISTANCE AND NOTED THE BLACK SAIL. IN HIS MISERY HE KILLED HIMSELF BY JUMPING OFF THE CLIFF.

Sir Arthur Evans

The story on the last page is a legend – some would say a fairy story. However, many legends have some truthful parts. The question was, did this one contain any real history? No one knew until the early years of this century, when a man named Arthur Evans was responsible for the digging up of evidence for a totally new and hitherto unknown civilisation. The Minotaur legend could then be given an historical background.

Evans was born on July 8th, 1851 at Nash Mills in Hertfordshire. His father was a well known collector of ancient objects. Arthur was sent to Harrow and then to Oxford.

His adventures gave him material for the articles he wrote in the English newspapers. He was actually in England in 1878 and seized the chance of seeing the treasures dug up at Troy by Heinrich Schliemann. Coins and seals in an unknown language drew him to Crete in 1894 where he met a local archaeologist who had explored a little in the area of Knossos.

He spent several years digging there himself. To make sure there were no legal problems he bought the land. There was no longer any need to get permission for his excavations.

Sir Arthur Evans rebuilding the palace at Knossos

Sir Arthur Evans at Knossos

From his late teens onward, he was a keen traveller and loved nothing better than a tour of the less familiar parts of the world. He didn't mind hardships a bit if he could escape to places that were off the beaten track. In 1875 he was in what we now call Jugoslavia, when the native peoples revolted against their Turkish rulers.

What came to light was no less than a new European civilisation, able to take its place alongside those of Iraq and Egypt. In all, there were over 5½ acres of the ruins of a colossal palace with hundreds of rooms and apartments.

What Arthur Evans found you can read about on the following pages. His discoveries made him famous. He had already been appointed curator of the Ashmolean Museum at Oxford in 1884, a position he was to hold for a quarter of a century. He also became Extraordinary Professor of prehistoric archaeology at Oxford University in 1909, after retiring from the museum.

He had uncovered a 4000 year old civilisation and was one of the first to say that here was a corner of Europe with a written language dating back almost as far.

He was knighted in 1911 and died only three days after his 90th birthday in 1941. What he made of the Theseus legend you can see in the next few pages.

Section 1 *The Minoans*

The palace at Knossos

Let's ask one of Sir Arthur Evans's men what it was like to dig up the palace at Knossos.

'It was a long time ago but probably one of the most important "digs" ever done,' he says. 'The remains of walls were only a short way down in the ground, and over the years we found a palace covering nearly six acres, surrounded by a somewhat larger town.

'The palace consisted of over a thousand rooms on at least five different levels ranged around a central courtyard.'

'What sorts of rooms?'

'All kinds, because the palace was used for varied activities. It was a house for a king, a court of justice, a suite of offices; it had reception, dressing, bed and dining rooms, a school, chapels, granaries, storerooms, studios, dungeons, annexes, workshops and many others. The most interesting, in my opinion were the bathrooms with their sunken baths but I suppose that the most important was the so-called throne room.

'The throne was made of stone and imitated the remains of a wooden one which we found in another part of the palace. On either side of it were long stone benches set against plastered walls. On the walls were painted plants and imaginary animals, reckoned to be griffins. That was the one thing about the palace that amazed everyone – the frescoes.'

'Frescoes?'

'Yes, the pictures painted on the plaster walls. You can tell from them what the Minoans looked like – medium sized, with long, dark hair and large, dark eyes. The common form of dress was a wrap-around kilt or loin cloth. Against that, we did find a statuette – of a priestess, I suppose she was, holding a snake in each hand. Now *she* was wearing a floor length skirt and a frontless blouse or jacket.'

'You said "Minoans": how do you know what they were called? Did they have a written language?'

'Yes, they had at least three, and only one of them has ever been deciphered. But we don't know what they called themselves. Evans got the terms "Minoans" from "Minos", the king in the legend.

'Another interesting word was "labrys" which means "double axe". This symbol turned up all over the

The throne room

Plan of the palace

10

place, in metal and in stone. It may have been a religious sign or a sort of royal trade mark. The word "labrys" is supposed to have given rise to the word "labyrinth".'

'That means "maze", doesn't it? Did you find any traces of the maze – you know, the Minotaur's lair?'

'I've a theory that the palace itself was the maze – the enormous number of rooms must have caused no end of confusion, especially to strangers. Can you imagine being given directions to a distant room? "Go along this corridor, turn right, second on the left, then turn right, down the stairs, take the third turn on the right through the little courtyard, up the stairs—". Wouldn't you tell everybody it was a maze?

'The excavators thought it was more like a jigsaw puzzle than a maze – a lot of it had collapsed on itself. Sir Arthur had to reconstruct a good deal of what was found first before we could go any lower.

'I think what he rebuilt gives a marvellous idea of what the palace might have looked like. Notice, I only say, "*might* have looked like." Some people say he made it seem like a Victorian hotel and that he hadn't any real authority and not much evidence for what he did.'

'Do you think that?'

'Heavens, no. I believe his reconstructions were absolutely vital. No one could have any idea of what Knossos was about if he'd just left the ruins as he uncovered them.

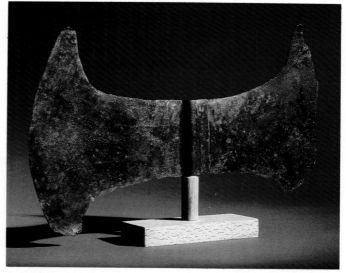
A double axe

'Just a couple of other items. Some walls seem to have been designed to withstand earthquakes, one of which could account for the so-called bellowing of the Minotaur. Survivors of severe tremors often mention a continuous earsplitting roaring that they heard. And other walls show signs of the various burnings which destroyed the whole complex on more than one occasion. The last time was probably in 1375 B.C. The fire marks on the west facade of the new palace show that there was a southerly wind blowing at that time all those years ago.'

Earthquakes and fires caused the palace to collapse.

Daily life

The old Cretan road

The palace at Knossos had a great many places for storing farm produce – large pits, sometimes lined with lead, and rows of enormous clay jars. These were obviously used to contain the taxes from the peasants – paid in olives, olive-oil, wine and different kinds of cereals.

In common with other ancient civilisations, that of Crete depended on farming. The farmers had only a few horses but there were cattle, sheep, goats and pigs. Barley and wheat were grown, as were olives, grapes and almonds – in small, terraced fields. Ploughs and carts were drawn by oxen, probably under the control of slaves. Anyone could own a slave. If a slave were lucky, he might be taught his owner's trade.

Apart from ox wagons, goods could be carried on a pack donkey or a slave's back. There were tracks and paths all over Crete and even a few stone-surfaced or 'metalled' roads. One of them, near the palace itself, is reckoned to date from about 1500 B.C. and is thus the oldest metalled road yet discovered. Other forms of transport included going by boat along the coast and travelling (at least for palace officials and servants) in litters.

Farmers lived in little square, lime-washed houses in small towns or villages. Apart from Knossos, there were over a hundred other large and small towns. Sometimes an official had his office and living quarters at a distance from the palace and a village or town would grow up round it.

Houses for the well-to-do, particularly those near the palace, were made of sun-dried bricks on a burnt brick or stone foundation. Often, they were two or three storeys high. Lower floors were of stone, cobbles, cement or beaten earth, with sheepskin rugs to keep bare feet warm. Some outside walls were faced with square stones. There were the usual doors and windows (although earlier and poorer houses had neither: if you wanted to get out you climbed a ladder to the top of the wall, hauled the ladder up after you and used it to get down to the well drained and paved street. Someone still at home pulled the ladder back inside!).

Later doors were of wood and windows were divided into two or three with stone bars. The spaces were filled with oiled and stretched animal skin. Most houses had flat roofs for summer sleeping. Occasionally, there were storage cellars under the houses of merchants, of which there were quite a few in the ports, together with the houses of ship builders, chandlers and fishermen. Fishing boats must have been kept busy. Food from the sea was very popular if the pictures of various fish, octopuses and dolphins on Minoan pottery are anything to go by.

None of the towns seem to have been protected by

A merchant's house at Knossos

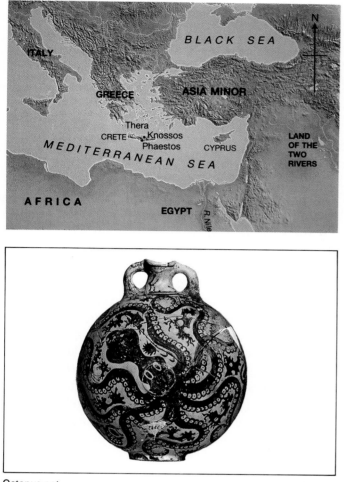

Octopus pot

Wall painting

walls – the Minoans relied on the fact that they were on an island so that their enemies would be dealt with by Minoan ships. They had a large fleet which traded widely all over the eastern Mediterranean.

Apart from the houses mentioned above, there were middle-sized dwellings occupied by gold and silver-smiths, weavers, carpenters, potters, basket makers and other craftsmen. As well as the two- and three-storied houses, there were villas with rooms round a central courtyard. Walls were often plastered on the inside and painted – red was a favourite colour. On these hung china plaques of fish or animals, rather like the flights of pottery ducks in some homes today.

The houses of the better off might have a dozen or fifteen rooms. Those with more than one floor had an outside staircase leading to the roof. Bathrooms were not unknown. The queen's bathroom in the palace had a terracotta bath which the servants had to fill from jugs of hot water. Afterwards, the used water was baled out and poured away down a drain in the same room. Richer Minoans had bathrooms of the same type.

Rainwater was used to flush the lavatory which often had a wooden seat. Water running off the roof was channelled into stone conduits which also carried waste away from all floors into a main drain. Heating in the cold weather might be provided by a portable metal brazier which burned charcoal.

Work was not the only way of spending your time. You could take part, or just watch or listen to a variety of pastimes including conversation, tumbling, wrestling and boxing, bull-leaping (see page 16), board games or dancing. Another interest must have been fashion but we only know how the people dressed from statuettes and wall paintings. The last named show women as white-skinned and men as brown. Whether this was like life or just the way they painted we don't know. It may suggest that women stayed indoors while men lived an outdoor life and were sunburnt.

Both men and women are incredibly wasp-waisted, the males with broad belts round their middles. Ladies wore their hair piled up high and kept in place with a bandeau. Both sexes were very fond of jewellery, wearing anklets, bracelets, necklaces and earrings of gold.

13

Religion

A goddess begging a favour of Zeus

It could be that a number of gods and goddesses whom we think of as purely Greek may have started off in Crete. For instance, Zeus (later taken over as Jupiter by the Romans) according to legend, spent his childhood on Crete. However, gods often took second place in the Minoan mind to goddesses. Cretans spent more time worshipping a mother goddess than almost any other ancient people – with the possible exception of the Maltese of those days.

On Malta there are many fine stone-built temples associated with statues of a mother figure. They are more than 5000 years old but we know very little of Maltese prehistoric religion or beliefs. Of course, we know little more of what Cretans believed; we can only put forward guesses based on the things they left behind.

To start with, there are no huge elaborate temples such as the ones in Egypt or Mesopotamia. Nor are there any gigantic statues of divinities. There were instead shrines of various kinds, many of which were adorned with the sacred bull's horns or the equally sacred double axe. The shrine might be in a palace (there are several shrines at Knossos itself) alongside a track, in a cave or on top of a mountain.

In earlier civilisations, religion was entwined with warfare, sometimes accompanied with what seems like senseless cruelty to captives. Cretan art so far discovered shows no sign of glorifying fighting, military leaders, or even hunting, even though we do know that some hunting took place.

Mother goddess worshippers washed on entering her shrine or were annointed with oil – probably to wash away their sins before being shown into the holy presence. Occasionally, the priestess would blow a blast on a sea-shell horn to summon the goddess.

The worshipper then deposited the gifts or offerings he or she had brought. These might be food and wine or something more valuable – a necklace, a swordblade, a gold or pottery double axe or an armband. One shrine had two stone chests let into the floor, perhaps to hold the offerings.

A Minoan shrine

Cretan funeral rites

In one wall painting, women are shown praying and pouring wine as a libation to one of the pillars which held up the shrine roof. In another part of the palace at Knossos, towards the north west corner is an area flanked by two shallow flights of stairs at right angles to each other. We have no proof, but it may be that this was a theatre for religious dances. Some statuettes of dancers also show poppy heads and it's possible that the performers were drugged with opium. Possibly they didn't realise the terrible danger of taking this substance.

Burial customs, like most other religious matters on Crete, have to be guessed from very small pieces of evidence. We know that tombs shaped like old-fashioned beehives were being used as early as 2800 B.C. and that this particular type remained fashionable for almost a thousand years.

The dead bodies were laid above ground inside the building and decked in all their finery. The commonest kind of coffin was made of clay but one example was cut from limestone and coated with white plaster. Its four sides are painted with pictures almost certainly dealing with the funeral. On one long side, mourners are bringing a small boat and a pair of young bulls to be sacrificed. The dead person is shown as alive and accepting the offerings. Women pour wine into a special vase surmounted by double axes. A man in the background plays his lyre, a kind of harp. The second long side shows the sacrifice, with music provided by a double-pipe player. The short ends of the coffin show the dead person and the goddess in a chariot. One painted chariot is drawn by goats and the other by a pair of the imaginary animals called griffins.

Hardly any of the Minoan tombs have been found undamaged – many have been robbed and totally destroyed. However, not far from Knossos, at a place known as Arkhanes, a Minoan woman's body was discovered. It was dressed in a gold trimmed gown and was wearing a good deal of gold jewellery – rings, necklaces and a pair of tiny gold boxes lying on her chest.

One of the rings has a picture of a religious rite, showing the great goddess, in a similar costume to the dead lady, conducting a ceremony aimed at encouraging plants and farm crops to grow. There are some butterflies round her shrine and people dancing.

Bull leaping

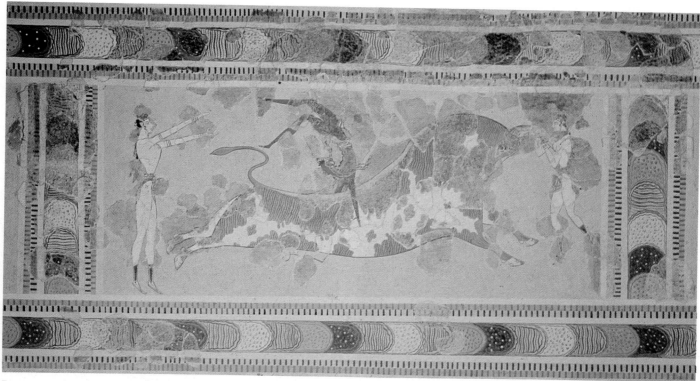

Fresco of bull-leaping from Knossos

Catching a wild bull, on the cup of Vapheio

One of the commonest souvenirs you can buy on Crete is a plaque, usually of baked clay, showing a youth holding the horns of a galloping bull and vaulting over its back. We know the bull was worshipped: even a stylised carving of its horns was looked on as sacred. The origin of this modern tourist's keepsake is to be found on seals and in a wall painting of the same kind of gymnastics.

It seems that bull leaping was either a religious or a popular performance rather like the 'games' that went on in Roman amphitheatres, or perhaps like bull-fighting in modern Spain, except that the bull was not killed.

The animals were apparently hunted and caught alive. There were many ways of trapping wild bulls. The two most popular seem to have been tripping the animals with ropes and tying them up, or driving them towards a large net stretched between two stout trees. Occasionally a cow was tied up nearby to act as a decoy. Someone had to wait until a suitable beast was entangled, jump on its back and wrestle it to the ground. Then the helpers would hobble its hoofs and take it off to an enclosure near the arena. Once there, it would be broken and trained.

Some modern experts believe that what is shown in the painting – the acrobat grasping the bull's horns and then turning a somersault over its back – is quite impossible. However, if the animal was tethered, tamed or even drugged, it might not have been.

This is not to say that there was no danger. The team of acrobats must have been at risk all the time – a false move, bad timing or a much too frisky animal and there was always the chance of one of the leapers being gored by a bull's horn. We mustn't exaggerate the danger; after all, performers on the high trapeze in a modern circus are in just as much peril but accidents are fortunately rare.

Rare or not, accidents did occur, even when the nearly naked leapers had trained intensively for months. Thus a steady stream of recruits was needed. Almost certainly this state of affairs gave rise to the part of the Theseus legend which spoke of seven Athenian boys (and the same number of girls) being sent to feed a bull-headed monster on Crete. Might this have been a half remembered attempt to describe the continuous tribute of young people to train as bull leapers for king Minos?

Michael Ventris and Linear 'B'

This is the famous Phaistos Disc. Phaistos (together with Mallia and Zakros) was one of the palaces on the island of Crete and was excavated by Italian archaeologists. The disc was dug up there. It is a little less than a foot across, made of clay and is stamped with 45 different characters.

Although it has been known for many years, its message is a complete mystery. No one has been able to decipher it. But that has not stopped both the expert and the not so expert from guessing its purpose. 'A set of mathematical tables', 'a peace treaty' and even 'a piece of music' were some of the suggestions.

Apart from the disc, other kinds of writing are known to date from the Minoan period. It took some time before any progress was made. The writings that survived were done on clay tablets which were often baked hard in accidental palace fires. Perhaps the Minoans also wrote on leaves, wood panels and other perishable surfaces.

The two most important methods of recording facts both had characters drawn with straight lines. Evans called them 'linear' because of these line-drawn 'let-ters'. The two different systems were called 'Linear A' and 'Linear B'.

In 1936, Sir Arthur gave a lecture in London dealing with his discoveries in Minoan Crete. He showed slides of the various clay tablets and explained that the languages were unknown and had defied all attempts to translate them.

In the audience was a fourteen year old schoolboy named Michael Ventris. He was so interested that he vowed then and there that he would one day disentangle the meaning of the clay tablets. When he grew up he joined the Royal Air Force, for the Second World War had started just after his seventeenth birthday. When the war was over, he trained as an architect but he never lost his love or enthusiasm for the unknown tongues of ancient Crete.

He went to work on them, trying various different ways to break through. Some of his methods were those used to decipher enemy codes in wartime. One of the first things to do is count the number of 'letters' or signs to see how often each one is used. If the number is fairly low – somewhere between, say 12 and 50, the language is probably written with an alphabet, just as English is. If it runs into several hundred, it's probably a syllable system. This means that words are split into parts or syllables and each one has a separate sign. This is difficult to do in English but less so in some foreign and ancient tongues.

To make it more difficult, there were occasionally little drawings of the thing being written about and even crude attempts to spell the word. A further obstacle for Michael Ventris was that no one knew what language 'Linear B' was written in. The only thing the wise men were sure of was that it was about 600 years too early to be Greek. Ventris himself thought that it might be the language of the Etruscans, a people who flourished in Italy before the coming of the Romans.

After some incredibly complicated work on the clay tablets of Sir Arthur Evans, Michael found what he thought were place names and then – following a little drawing of a three legged support for a cooking pot – the signs for TI-RI-PO-DE (Tripod).

TI – RI – PO – DE

Using the signs discovered, he set them against similar signs on the tablets and began to realise that these were lists of stores, accounts of taxes and tribute paid, names and numbers of soldiers and similar matters. Furthermore, he was extremely surprised to find that the tablets were written in a kind of Greek – the one language everyone had dismissed as impossible.

Although some people still do not accept this solution even now, there has never been a better answer put forward. Ventris was tragically prevented from doing further work by a fatal car accident. He was only in his early thirties.

Perhaps you'd like to see what code breaking is like. Try your hand at this: 'BLLG BL CG GWKLL EISEIQ EV GWL WLCGW. TSLCUL HL GWLKL CVF FEVG HL SCGL. GWL VLYG GWKLL MLLQU MRSS ULL GWL LVF EP GWL BCGGLK.'

1   Count the different signs. If there are hundreds, it's probably a syllabic language: if only (say) 12–50 characters, it's alphabetic. (To make it easier this one's in English.)

2   Which letter occurs most often? (In English it's 'E'.)

3   Put 'E' wherever the most frequent letter sign comes.

4   Can you now guess which three signs are the word 'THE'? (It's the commonest '3-group' in English.)

5   When you are sure of (4), put 'T' and 'H' in their proper places. This should suggest some other words – for instance, there are several words which begin with the combination 'THE' – e.g. 'THERE', 'THEIR', 'THESE', 'THEN' etc.

6   By now you'll have enough letters to solve the code message.

A Minoan scribe

Atlantis

Plato

A Greek writer named Plato wrote a book called *Timaeus*, in which he describes how some priests of ancient Egyptian gods had told a Greek tourist called Solon a very curious story.

Until about 9000 years ago, he was told, there existed a huge island, which was rich and powerful and whose ships and soldiers had conquered all the lands around the Mediterranean. Only the Greek city of Athens had resisted them.

The people of this island had become so proud and conceited that the gods were offended and they sent tidal waves and earthquakes to punish them. In a single night, the island was overwhelmed and sank beneath the waters of the ocean.

Solon was told that the island was in the Atlantic beyond the Pillars of Hercules (the Straits of Gibraltar) and was known as 'Atlantis'. Modern scientists have been completely unable to find any evidence to support this remarkable legend and some have wondered if Plato didn't make the whole thing up.

Did Atlantis look like this?

Romantic novelists and writers of adventure stories took up the legend enthusiastically and produced endless tales of the miraculous Atlanteans. More serious thinkers began to wonder if there was, at least some truth in the story. Perhaps the timing or location of the disaster were wrong.

Then in the 1960s remains of a brilliant civilisation began to turn up on the island of Santorini, or Thera, to the north of Crete. Thera is almost a third of the way between Knossos and Athens on the Greek mainland.

From about 1967 onward, rain washing down had revealed traces of wall paintings very much like the ones in the palace at Knossos. Excavation confirmed that this indeed was a palace similar to that of king Minos on Crete.

There were also streets lined with houses of the same type as those found near Knossos, some as high as three stories. The archaeologists found rooms with giant storage jars and several plaster panels painted with scenes of everyday life.

One painting shows some beautifully graceful antelopes: another, a young fisherman with a string of fish in each hand. This is one of two boy boxers. They seem to be about ten or eleven years old and are wearing nothing but a waistband and what appears to be a boxing glove on the right hand only. A large part of this fresco is a restoration and if the modern artist is correct, it's very difficult to imagine what sort of boxing is being done. One suggestion is that each boy holds the other's hair with his left hand and hits with his right! This is not so unusual as might be supposed: not long ago in England, fighters stood still, toe to toe and took it in turn to hit each other.

From the kinds of soil covering the ruins of the palace and houses, archaeologists were able to say roughly when and how this rich and varied Minoan type civilisation had come to an end.

There had been a series of massive volcanic eruptions at about 1500 B.C. which had virtually blown away most of the island. The first eruption had blasted a hole in the side of the old crater, sea water had poured in onto the white hot lava and a titanic explosion was the result. This bang may have been the loudest ever heard. In the far east, a hundred years ago, an island called Krakatoa had exploded and the noise had been heard over a thousand miles away.

Scientists say that the disaster on Santorini was a good deal worse and the bang very much louder. There must have been huge tidal waves and it is tempting to think that the end of Minoan civilisation on Crete may have been the result of gigantic earthquakes and volcanic eruptions. These might have given rise to both

the Plato story of Atlantis's destruction and also the legend of the Minotaur's death.

On the other hand, the truth (if we ever come to know it) may be nothing like the ideas given above!

Minoan fisherman, and boy boxers (below)

Homer

The royal palace at Pylos at the time of the Trojan war.

One of the Greek kings in Homer's poems is called Nestor. Nestor's palace was at Pylos in the South of Greece. In recent years the palace has been excavated. Here you can see an artist's reconstruction of what it would have been like. The detail shows that Nestor's bathroom was not very different from a modern one, though there are no taps with hot or cold water. Odysseus's palace must have been very like this one.

It is 730 years before the birth of Christ. We are waiting in the hall of one of the Greek kings. The meal is over – the last olive and sliver of cheese have been eaten – and boys are serving the wine. We ask our neighbour what kind of entertainment we are to be given. He replies, 'Why, it's Homer!'

We ought to be impressed but our friend can see that we've never heard of him, so he explains.

'Homer is blind, you know: he has to be led about everywhere. He was a slave many years ago but he became famous for his recitations. Whilst serving his master, he found he could repeat a poem of any length even if he'd only heard it once. He has friends who've written down the poems he's composed. Among the longer ones is a story about part of the Trojan war.'

There is a stir and Homer is led in by his attendants and settled down. His lyre is handed to him and he begins to tune it by tightening or loosening the pegs. He clears his throat, strikes a chord and starts to recite.

The poem is one of the longer ones of his own composition. At least the form of the thing is Homer's although the actual story has been passed on down the years for about five centuries before Homer's time. It is all about the adventures of Odysseus on his way back home to Greece after the siege of Troy. Now that it has been set down on paper, future generations will also be able to enjoy the Odyssey, as it is called.

Odysseus, or Ulysses, as the Romans knew him, set out from Asia to sail back to his queen in Ithaca, Greece. Unfortunately, he and his men were lured into one trap after another and many of them perished.

They were driven by storms to the land of the lotus eaters, where anyone who consumed the lotus forgot where he came from or even who he was. Then they were captured by a Cyclops, a one-eyed giant named Polyphemus who kept them prisoner in his cave, together with a flock of huge sheep. Odysseus had told the giant his name was 'Noman'. When the Greeks seized their chance to escape by putting out the giant's eye, Polyphemus shouted and roared until his neighbours came to ask what the trouble was. 'Noman has attacked me!' screamed the giant.

His neighbours then called back from the other side of the boulder which sealed the cave mouth, 'If no man is hurting you, it must be a nighmare. Go to sleep.' And they went away.

In the morning the giant tried to stop the Greeks getting away by riding out on his enormous sheep. As the animals passed through the rock opening, Polyphemus ran his hands over their backs. He could no longer see and wasn't aware that Odysseus and his crew were clinging to the wool of each sheep's belly.

Later, one of the gods gave Odysseus a goatskin bag, which he said contained favourable winds to blow them home. Some of the crew thought there must be treasure in the bag and undid it. Out rushed the angry winds and once more the ship was sent off course.

The crew escaped from the rock and whirlpool of Scylla and Charybdis and from the island of sirens, who would have entranced the sailors by their songs, had they not sealed their ears with beeswax. By a trick, the leader managed to free his men from a spell cast by a witch named Circe, which had turned them into pigs.

Finally the wanderer returned to his homeland of Ithaca in disguise. He found that his wife Penelope had been besieged by men wanting to marry her and thus become king, firmly believing that her husband must be dead. Penelope still thought he would return even after ten years of wanderings, so she said she would marry the man who could bend her husband's great war bow – knowing that it was unlikely that anyone could do so.

She was right. None succeeded until the lonely, dust-covered stranger not only managed to bend the bow, he also slew the pestering suitors with sharp arrows.

The notes of the lyre die away and we ask our friend about the other legend he mentioned – the siege of Troy.

'Come in a week's time,' he says, 'Homer will recite that story then.'

Odysseus slays the suitors

The Trojan War

Once again we sip our wine and listen to Homer as he tells us something of what happened when the Greeks fought the Trojans. Unfortunately, Homer doesn't deal with the whole story, so we have to ask our companion about the rest. From the two of them we can piece together most of the main events.

Troy, or Ilium, as the Greeks called it, was a city in the country that we modern people call Turkey. It stood on the Mediterranean coast, a short distance from the sea. Its king was an elderly man named Priam. He sent his son, Paris, as an ambassador to the court of Menelaus, king of Sparta.

Tragically for all concerned, Paris fell in love with Helen, the beautiful wife of Menelaus. He took her away secretly to his ship and sailed back to Troy.

Menelaus was beside himself with fury and called on his fellow Greek kings to rescue the lady and help him to his revenge on Paris. Those who heeded his call and turned up with ships and men were the aged Nestor, Diomedes, Odysseus and Menelaus's brother Agamemnon, king of Mycenae, who became the expedition leader, and Achilles, greatest of the heroes.

The fleet sailed across the seas and dropped anchor off the coast near Troy. The city was a formidable sight with its 25 feet high walls. The Greeks made camp and thought about the problem. All they could do was to encircle the city. Unluckily for the besiegers, there were countless sorties by Trojan heroes such as Hector, Paris and Aeneas. These drove off the enemy and raised the siege – even if only temporarily.

Often, the encounters took the form of single combat. Achilles, however, had quarrelled with Agamemnon and went off to sulk in his tent. His closest friend, Patroclus, borrowed his armour and took his place. Sad to say, he was slain by Hector, the Trojan commander.

Achilles drags Hector round the walls of Troy

Achilles then sought out the slayer and avenged his friend's death, only to die later himself, killed by a poisoned arrow fired by Paris. The arrow had struck Achilles in the heel, the only part of his body that was not magically protected.

Homer's recital finishes here so we have to get the rest of the story from our friend. He tells us that the siege dragged on for another ten years. Then the Greeks had a brilliant idea. They made a huge wooden statue of a horse and left it in front of the city. Then they set sail and their fleet headed out to sea.

The next morning, the astounded Trojans saw this enormous horse and a bay empty of Greeks. 'The enemy has gone!' they shouted, 'and they've left us a present.' Some wanted to bring the horse into the city but others were more cautious, saying, 'Beware of the Greeks when they bring gifts.'

However, there didn't seem to be any danger, so the horse was brought inside the walls and the Trojans began to celebrate the end of the war. By the evening,

The wooden horse is dragged towards Troy

most Trojans had drunk more than was good for them and a good many collapsed in a stupor, not knowing that they were about to lose the war. They were ignorant of the fact that some soldiers were concealed in the hollow body of the horse.

In the middle of the night, the Greek warriors unbolted a trap-door in the animal's body and shinned down ropes to the ground. From there they went around, unchallenged, opening the city gates.

The main Greek army, which had not gone home, but merely hidden in the next bay, swarmed in through the gateways. Nearly all the men and boys were slain. The women were taken back to Greece as slaves. Priam was dead and Troy utterly destroyed. Helen was returned to her husband, Menelaus, and the war was over.

We have imagined what it might have been like to listen to Homer speaking the lines of his famous poems, 'The Odyssey' and 'The Iliad'. But it is imagination, nothing more. In fact, no one knows exactly when Homer lived. We're not even sure if he was one man or several and many towns in the Mediterranean area have claimed to be his birthplace.

On the next page, in the story of Heinrich Schliemann, we'll try to see if *this* legend has any basis in fact.

Heinrich Schliemann

One man who was sure that Troy really existed and that the Trojan war had actually been fought was Heinrich Schliemann, a somewhat unlikely archaeologist, who spent a good part of his life making money.

He was born in 1822, the son of a poor clergyman in Germany. When he was seven, one of his Christmas presents was a book of Greek legends. At that time, Troy had vanished so utterly that no trace of it still existed and no one had the faintest idea where it had once been. Young Heinrich promised his father that one day he would find the site of Troy and prove it to the whole world.

He was brought up amid grinding poverty and was forced to find ill-paid work as a grocer's boy at the age of fourteen – not the background you'd expect for a great excavator. Ill health made him give up this job and then he became a cabin boy on a ship which was wrecked off the coast of Holland.

Heinrich Schliemann

An artist's impression of Troy

In Amsterdam he was employed in a merchant's office and began to learn foreign languages – Dutch, French, English, Spanish and several others, taking no more than six weeks over some of them. In 1846 he went to Russia for his firm and a few years later to America where he made a fortune buying and selling.

He travelled more and more extensively and had made so much money by the age of 41, he was able to retire from business and devote himself to archaeology. By the late 1800s he was convinced (from details in the Homeric poem) that Troy lay beneath the soil of a hill called Hissarlik, in modern Turkey.

Between 1870 and 1873 he excavated Hissarlik with over a hundred workmen. They toiled hard and removed mountains of soil. The finds consisted largely of pottery fragments – but pottery of a type not previously known. The diggers also found walls and even walls underneath the walls. Schliemann was forced to conclude that one city had been built right on top of another, and earlier, one.

Nowadays, we know that when a city was sacked or burnt down, the rebuilders often just levelled off the ruins before starting again. Schliemann uncovered the layers of no less than nine different occupation levels,

the earliest going back (we now know) to before 3000 B.C.

In those days (i.e. the 19th century), archaeology had scarcely begun as a science and Schliemann had little idea of the dates of the various layers. In the fire-scorched ruins of the second and third oldest cities, he found a tremendous hoard of valuable items. He immediately announced that he had found 'the treasure of king Priam'. In fact, the articles in question that he had found were a good thousand years earlier than the Troy of Homer's poem but Schliemann was dazzled by his boyhood dreams of finding the classical city.

The treasure consisted of thousands of different things – beakers, cups, plates, necklets and necklaces, bangles and bracelets. There were over 8000 gold rings.

Heinrich continued digging into the hill of Hissarlik on and off for the next few years. Other archaeologists came to help from time to time but many scientists were not convinced that this was indeed Homer's Troy. However, in our own time, most people are now sure that Schliemann really did find the ancient site of the Trojan war.

Remains of the Trojan walls

Schliemann at Mycenae

Schliemann mistakenly claimed this as Agamemnon's death mask.

We've already seen how Agamemnon became the leader in charge of the Greek army which attacked Troy. He was the king of Mycenae, which had been for some time the leading city of Greece. Thus it was natural for the 'King of Kings' to give orders to everyone else.

Unlike many other sites, Mycenae had always been visible to travellers throughout the ages. Ancient writers referred to it as a powerful and 'golden' city state. It is not, therefore, surprising that Schliemann should turn his attention to it. In 1874 he made his first survey, digging a number of holes and trenches. Nothing of great interest came to light but he discovered in some deep graves no less than five bodies.

The corpses were loaded down with gold. The faces were concealed beneath masks of gold and, buried with the dead, were weapons, cups, beakers and ornaments – all made of solid gold. Schliemann immediately jumped to a wrong conclusion, just as he had done at Troy. 'I have looked upon the face of Agamemnon', he said. However, these bodies were not of Agamemnon and his family; their funerals had taken place a good four centuries before the famous king of Mycenae was born.

Schliemann also turned up decorated daggers, bowls, plates, cups and beakers. Often the decoration took the form of illustrations of daily life, which in this case consisted largely of military and hunting scenes. Mycenaeans, it seemed, were far more warlike than the gentler Cretans.

Other archaeologists followed Schliemann's example and, over the years, a picture of the powerful and aggressive city state emerged. Outside the walls, graves even older than those first found were discovered. Again, the funerals were accompanied by many articles of silver and gold. Clay tablets were found in house foundations just outside this extremely strong fortress. The tablets were written in what we call 'Linear B' (see p. 18).

The main entrance is known as the 'Lion' gate because of the magnificent sculpture above the stone lintel. Apart from the size and strength of the city walls, attackers faced an even greater problem. Almost certainly unknown to would-be besiegers, the Mycenaeans had a secret water supply, within the defences, from an underground spring, just outside.

Then a number of 'tholos' tombs came to light. Each contained just one king and his immediate family together with the funeral offerings. The remarkable thing about these old-fashioned beehive shaped graves was their huge size. One tomb, known as the 'Treasury of Atreus' measured nearly 48 feet in diameter and one of the lintel stones over the doorway has been reckoned to weigh over 100 tons.

We know from written Greek history that Mycenae was the strongest and richest city in the whole of Greece. Its art and fashions, types of weapons and pottery were exported and copied all over the known world – to such an extent that it is possible to speak of other 'Mycenaean' cities in the Greek world.

It was Homer who used the word 'golden' to describe Mycenae: it was a fitting description. More gold has been found there than in all the other ancient Greek sites put together.

Mycenaean merchants exported wine, perfumes, manufactured metal goods plus clay pottery and they imported ores of both precious and base metals. They set up trading posts which developed into colonies in many places. Finally, the Mycenaeans (probably taking advantage of the chaos caused by the earthquakes and eruptions in the area) invaded Crete and captured Knossos. They ruled the island for more than two centuries.

Even if a Mycenaean man was a merchant, he would also have a farm of some kind on which he grew grain and fruit and raised sheep, goats, pigs and cattle.

At Mycenae, there were found, as well as the fortress, substantial foundations of houses belonging to the traders and to workmen.

All indications show that the city was powerful, rich and extremely well protected. It is therefore all the more surpising to find that the fortress of Mycenae was conquered by bands of wandering Greeks known as Dorians about eleven or twelve centuries before Christ.

The Lion Gate at Mycenae

Achaeans and Dorians

Armour of the period

With the end of the Trojan war, the rule of the Mycenaeans also came to an abrupt end. No one is sure why this happened: some say the climate began to change and much of Greece became a good deal drier. Earthquakes and disastrous fires have been blamed as well, but the most likely cause was a series of invasions.

The land of Greece probably had as many outsiders fighting their way in as anywhere else in Europe but the first group we can put a name to were the Achaeans. Homer in fact uses the term for all the Greeks who fought in the Trojan war, no matter from which city they came. Therefore from his point of view, the citizens of Mycenae and similar strongholds were also Achaeans.

The new invaders were known as Dorians because they were believed to come from a place called Doris in northern Greece. Their conquering of the Achaeans started apparently well before 1100 B.C. When it was over, the whole of Greek civilisation had taken several steps backward.

Strangely enough, their weapons, although crude, were better than the beautifully wrought swords and daggers of bronze the Achaeans used. The reason they were superior was that they were made of iron. Thus, the Dorians were the people who introduced iron-working to the land of Greece.

The words 'Greece' and 'Greeks', by the way, were not the names they gave themselves but Latin words, bestowed by Romans of a later generation. Greeks of the generation after Alexander the Great called themselves 'Hellenes' but it's very unlikely that those who lived at this early stage thought of themselves as 'Greeks' at all, nor of 'Greece' as a nation. The Dorians conquered places and settled there, regarding themselves merely as citizens of whichever city they lived in.

It seems that the newcomers spoke a kind of Greek, although they knew nothing of reading and writing. The Linear B script died out completely. When the 'dark ages', as some people call them, were over – about four centuries later – the Greeks had borrowed a proper alphabet from the Phoenicians with which to record their language.

The Dorian invasions were thorough, overrunning

most of the important towns. Athens seems to have been an exception. Either the northern conquerors never tried to capture it, or they did try and were unsuccessful. Their chief town in the newly won south was Sparta.

One result of the Dorian invasion of mainland Greece was that some Achaeans left in large numbers, searching for somewhere to live that was free of the hated Dorians. The islands around Greece provided the safety sought by such groups. Others migrated to the western coasts of what is now Turkey.

Two of these groups of Achaeans were known as Aeolians and Ionians. They founded Greek colonies in Asia Minor that lasted until just after the First World War, in this century.

By about 800 B.C. the dark ages were drawing to a close. The Greeks had a common language even though there were local dialects. The first written accounts of many popular myths and legends date from this time, although they may have been passed on by word of mouth for centuries before the Greek alphabet was invented.

At the same period, iron-working became much more common. The first Greek colony in the western Mediterranean was set up at Cumae in Italy. Trade was beginning to recover and there was a rebirth of culture, including art, literature and science. In short, the classical age of Greece was about to dawn – with a certain amount of tension between the two rival cities – Athens of the Achaeans and Sparta of the Dorians.

As a footnote, it is only fair to point out that some historians don't accept the story of Dorian invasions. They say that the differences between the various Greek settlements can be accounted for by the separate development resulting from the isolation of cities by ranges of mountainous country. We don't have sufficient evidence yet to be sure.

Battle scene

The emergence of cities

Ancient Greece never became a nation in the same way that France, Britain, or even modern Greece did. The reason was touched on at the end of the last page and can also be gathered from a glance at a map.

Greece is a stony, mountainous land, shaped rather like a hand with fingers stretched out into the Mediterranean sea. There were odd spots of fertile land but they were few and far between. They were separated from each other by arms of the sea or steep ridges of limestone mountains. The country was almost divided in half by the Gulf of Corinth and the whole area surrounded by countless islands, particularly in the Aegean sea to the east of the mainland. The entire area of Greece is smaller than some American states.

Because the regions where people could get a living were cut off from each other, they tended to develop in their own way and had little to do with their neighbour. This isn't surprising if a town's nearest neighbour was on the other side of a mountain. The paths between would have been difficult to use in the summer and virtually impossible in the winter. Only occasionally did the towns join together – for instance when the whole land was threatened by foreign invaders (and not always then!). Another exception was for the regularly held games.

Apart from these examples, Greek cities and towns often fought each other for economic reasons, occasionally out of fear or even jealousy. Each town or city state was called a 'polis', a word which turns up in modern English expressions such as 'metropolitan', 'political' and 'police'.

A polis was a single town set amidst farm land. Some, like Athens, were large. Most others were fairly small and there were very many of them. One guess says that there may have been 150 on the Greek mainland and ten times that number in the colonies and trading posts scattered all round the Mediterranean and Black seas.

Some city states were founded in wild countryside, others on the sites of more ancient settlements. In the latter case, the old palaces were destroyed or burnt, sometimes deliberately, and what had been the original city became the inner citadel or fortress in the new and enlarged area. Here the population could flock in time of danger.

Danger wasn't the only reason the citizens got together – they had to have meetings for the running of the town or city. The bronze age kings had gone and at first their place was taken at the city level by any man strong enough to reign by force. After a while, however, some city states developed a system of self government, replacing the rule of the tyrant with that of the people. The very word 'democracy' comes from the Greek for 'people's rule'.

Although the citizens in such a method of government could and did meet to decide policy and law, the only ones entitled to an opinion and a share in the law making were the free adult males. This effectively cut out women, foreigners and slaves, so it was not really true democracy as the modern world understands it.

The average town had houses and temples huddled in narrow, winding streets and both sorts of building seem to have developed from a common pattern. They were made of wood or brick, more rarely of stone and were squarish with a pitched roof, probably of thatch. The door was often protected with a small porch roof standing on two side columns, maybe tree trunks in their first form. The roads in which these houses and temples stood were normally of beaten earth, sometimes dressed with a layer of gravel.

This was the sort of town in which Greeks of the early classical period lived. The largest of all the city states was Athens.

Early Greek town

33

The early days of Athens

We've managed to obtain a guide to show us round Athens. The time is the end of the dark ages and the beginnings of classical Greece.

'Hello,' says the guide, 'my name is Kallias. My city is Athens. We are in a part of Greece called Attica. That's on the eastern side of the country, just above the line of the Gulf of Corinth.

'Athens, you may remember, was almost the only Greek city not conquered by the Dorians who flooded in from the north. What we did have was a flood of refugees from other parts of the country – all of them running away from the invaders. We were very overcrowded for a time.

'Then many of them migrated to Asia Minor and set up their own city states among the barbarians – you know, those who can't speak Greek and whose language sounds to us like "Bar-bar-bar"!

'You can see how large the city is. In fact it's the most extensive and thickly populated in the whole country. Sparta and Corinth are the next in size and importance. These walls run a very long way: over three miles with half a dozen gates.

'If we stand in the very centre of the city we can see, away to the north west, an open space or town square which we call the agora. It means "a plain", "a level, uncluttered area" or "assembly area", though it's very

difficult to keep the agora uncluttered. You'd be surprised at the number of people who want to put up buildings on the smallest patches of land round the edges. Of course, when we want to hold a meeting, all the market stalls have to be cleared away. But there are so many statues and monuments now that it's difficult to walk a straight line – at least along the borders of the agora.

'Now, if you look away to the south-east, what do you see?'

'A large flat-topped rock. A hundred feet high perhaps?'

'About that. What you are looking at is our Acropolis. If it weren't there, you'd be able to see right down to the south east corner of the walls. What am I saying? If it weren't there indeed! If it weren't there, it wouldn't be Athens. After all, the Acropolis is the original town.

'When cities began to develop, they outgrew their hilltops and spread out at the foot. "Acropolis" only means "highest city". I'm told that there were royal buildings there once. Now, of course, all you can see are temples. Just below the rock on the south side is our theatre. You can't see it from here but you can from the Acropolis of course.'

'But where is the Parthenon? Surely one of the finest buildings anywhere in the world. I don't see any sign of . . .' We break off. The forehead of Kallias is creased into a frown. 'I'm sorry,' he says, 'I don't understand what you are saying.'

Of course he doesn't and it isn't really our place to tell him that in the future his city will be involved in a war with Persia and that the strikingly beautiful Parthenon will be erected centuries hence – partly as a monument to those who fell defending Athens from foreign invaders.

The agora with the Acropolis in the distance

How the cities were ruled

If we were to ask a historian to give us details of the way the city states were governed, he would probably say something like this: "To tell the truth, we don't know very much on this subject. About Athens, for example, or Sparta, we have some knowledge but for most of the other places, our information is either very scanty or it doesn't exist at all.

'We know that, during the bronze age, there were places in the Greek world that were ruled by royal families but when the Dorians invaded in about 1200 B.C. nearly all that sort of thing was swept away.

'Strangely enough, it was only in Sparta, the chief Dorian settlement that the idea of kings remained. The Spartans even went so far as to have two of them at a time. Mind you, the Spartan kings were really just army commanders, but they were members of a sort of council of elders: there were thirty elders all told. They had to be over sixty years old before they could take office and were then elected for life.

'In addition there were magistrates to look after the day-to-day affairs of the city. They were called "Ephors" and there were five of them.

The Spartan 'kings' sitting in the council of elders

Athenians voting

The Areopagus

'It was different in Athens. This was probably the first place on earth to use a system called "democracy" (see p. 33). It didn't happen overnight but took many years to develop.

'In the earliest times, Athens was ruled by nine rulers or "archons" drawn from ranks of upper class Athenians. They were elected by an assembly of all citizens who owned a certain amount of property.'

'It doesn't sound like democracy as we know it.'

'No,' says the historian, 'it wasn't. I think you already know that there were certain people who had no say at all in public affairs: women, poor people, foreigners and slaves. Archons ruled for one year only and then retired to a sort of House of Lords which met at a spot called "Areopagus" (The Hill of the War God). To begin with, it had wide powers but these were slowly cut away by a number of Athenian statesmen such as Solon, Cleisthenes and Pericles.

'Although most people could read and write, we don't think that early laws were written down. Most people knew the main ones by heart – just as modern people can recall a proverb to fit any kind of happening. Mind you, men tended to disagree as to which law ought to be applied. Some of the laws probably contradicted each other – but then, so do our proverbs. Don't say you've never noticed that "He who hesitates is lost" and "Look before you leap" mean absolutely opposite things. So with Athenian laws, which must have needed tidying up.

'Draco, an archon in 621 B.C., was the first to codify the laws and write them down. Some of them had such severe punishments laid down, that we still use the word "Draconian" to refer to modern laws with harsh penalties.

'By the 5th century B.C., the Athenian Assembly met almost once a week and every male Athenian over 18 was entitled to be present and to speak for or against any proposal and then to vote on it.

'Anyone could make his way to the Pnyx, the hill to the west of the Acropolis, where the Assembly met and argued about taxes, temples, treaties, road building, war and peace, or anything else that arose.

'A lot of the actual work (in carrying out the wishes of the Assembly) was done by the "Boule", a council of 500 men, chosen at random to serve for one year only. Anyone could be a member of the Boule but never more than twice in a single lifetime.

'For nearly two centuries, Athens led the world with its democratic way of life, until the city was captured by the Macedonians in 322 B.C.'

Solon, Cleisthenes and Pericles

Solon

Pericles

'These were the three men who did most to bring democracy to ancient Athens,' says our historian, 'so it might be a good idea to find out something about them.

'Let's take Solon first. He was a magistrate or archon, elected in 594 B.C., with special powers to do something about the laws – poor people had complained about the harshness of the code of Draco and rich people complained that there was too much unrest and disturbance.

'Solon's first decision was to ease the birth qualification for membership of the Areopagus: from then on you didn't need to be born an aristocrat to be a member. Solon then did away with the savage custom of selling someone into slavery because of an unpaid debt. It was his idea to set up a court of appeal open to anyone who was dissatisfied with an ordinary court verdict.

'He decided to leave Athens at that point to see how the experiment would work in his absence. In fact things went rather well, although the rich complained that they had given away too much power and the poor maintained that they had got too little.

'Oddly enough, it wasn't under the rule of an elected official but during the "reign" of a tyrant, the self-appointed Peisistratus, that the common people became really free and the control of the land-owners was finally broken.

'Peisistratus seized power twice illegally – each time by a trick. The first time he pretended to be wounded

and acted as though he was being chased by a murderous enemy. He pleaded for an armed escort to protect himself and then used the men he was given to take over the government by force.

'He was quickly driven out but returned once more. This time he pretended that an ordinary (if somewhat large) woman, whom he had arrayed in golden armour, was really the goddess Athena. He arranged for her to drive into Athens in a chariot and to tell the people to trust Peisistratus. This they foolishly did and the tyrant returned to power for the next twenty years.

'When he eventually died, his two sons succeeded him. One died naturally and the other one was murdered. There might have been a successful revolt by the land-owners. Instead, there was a period of chaos lasting two years, after which order was restored by Cleisthenes who gave people back their freedom under the law.

'He rewrote the rules of government and broke the power of the aristocrats by redrawing the boundaries of the regions occupied by each land-owning clan. In this way, the areas of the old ruling classes were broken up into bits and pieces of land, scattered here and there throughout the entire land.

'Another invention of Cleisthenes was that of 'ostracism'. If a man had behaved so badly that the city would be better off without him, the citizens could write his name on a broken piece of pottery (the Greek for which is "ostracon") and put it into a large container. If a man received more than 6000 "votes" he was banished for ten years.

Pottery used in ostracism

'Another famous statesman was Pericles, the great nephew of Cleisthenes. By his time (after the long and damaging Persian wars) democracy was as complete as it ever would be in Athens. Pericles presided over what has come to be looked on as a golden age.

'Under him, Athens blossomed as the capital, in all but name, of Greek art, liberty, architecture, sculpture, drama, poetry, history, philosophy, science, politics and many other human activities.

'Athens had become almost an empire, controlling the whole of the Aegean and much of the Greek mainland. Her allies paid her money to deal with their possible enemies. The Athenian navy protected them and Pericles used whatever money was left over from ship building to beautify his city. It is from this period that the superb entrance gates to the Acropolis were set in position and the world famous Parthenon erected on the rock's flat top.

'As far as public life was concerned, Pericles said that the law should be fair for all and that men should be honoured for what they had done, not for who they were. Violence was to be avoided and men must learn to give and take.

'By now the ancient ruling council, the Areopagus, still packed with aged archons, no longer had any powers beyond that of trying murder cases. The running of the city and its surroundings was the responsibility of the Assembly whose meetings anyone could attend.

'The command of the army and navy was entrusted to ten generals elected each year: these were posts of far too great importance to be left to a chance ballot. In fact Pericles himself continued to serve as an army general almost until the day he died in 429 B.C.'

Peisistratus and the pretended Athena

Everyday life

Harvesting grapes and olives

Many Athenians got their living from the soil. The richest had their farms nearest the city so that they could actually live in town houses, leaving their slaves and employees to do the hard work on the land. Their day consisted of rising from a couch, washing, breakfasting on a lump of bread dipped in olive oil or wine and water, and starting out on a tour of inspection.

Depending on the time of year, farm labourers would be breaking the soil with iron-shod ploughs, harrowing, broadcasting barley seed, hoeing, weeding or reaping. Attica, the area around Athens, was not all that suitable for the raising of cattle or the growing of wheat, so most farms grew barley, beans, lentils, grape vines and olive trees.

A man who had no need to supervise a farm would base his life entirely on his city house. This probably had rooms leading off an open courtyard. The house itself might have a stone foundation on which stood the timber framework and sun-dried clay brick walls. A porch roof supported on columns led into the main part of the house. This doorway was the only opening on to the narrow street. What windows there were mostly faced on to the central courtyard. bedrooms were behind the courtyard or even on an upper storey.

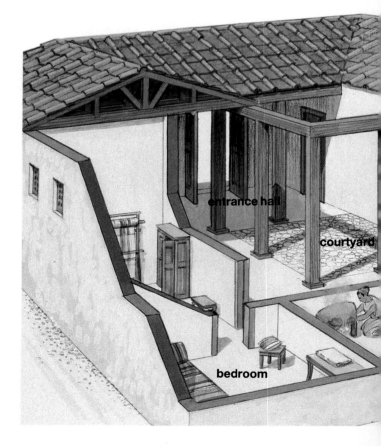

entrance hall

courtyard

bedroom

The roof was much less steep than those in modern Britain, for rainfall was seasonal and fairly low. What rain there was could be collected in a large stone cistern. The roof was made waterproof with thatch or clay tiles.

There were no separate pavements specially for walkers around the buildings, but the roads themselves were sometimes paved or gravelled. Gangs of slaves swept the streets, being careful not to tip any rubbish into the drainage ditches.

Once inside the house, the master would bolt the main door. His personal possessions were mostly kept in cupboards and his stores in well-built rooms, both of which could be locked. There was not very much furniture – a few small, three-legged tables together with some chairs and stools for the women and children to use at meal times. The master himself lay on a long low couch to dine. However, most families had a fine collection of painted pottery (see p. 44).

If the master had to go out, he would take down his cloak from where it had been hanging on a nail in the annexe to the main reception room. Other items similarly hanging on nails might include his lyre (or small harp), his strigil (or bathtime body scraper), together with his military equipment – sword, spear, shield and leg guards. All Athenians had to do military

Coins of the period

service and the richer a man was, the more equipment he had to supply. A poor man might only be able to afford a spear, whereas a rich one had to supply a riding horse (a war chariot in earlier times).

There were occasional housing shortages, particularly in war time when house building had to stop. A citizen could buy or rent a house in more peaceful times but he might, like Diogenes, sleep in a huge, broken and discarded wine jar.

The master, then, goes down to the agora, or public square, often used as a meeting place, but mostly given over to market stalls. Maybe he is doing the shopping. By now Athens is minting its own coins, which makes such expeditions easier. A slave walks behind carrying a basket for the purchases.

Many of the errands can be done at the wooden stalls standing in rows across the agora – items such as meat, fish, bread, poultry, fruit, vegetables, herbs, olive oil and flowers.

Other goods and services can be found in the rows of permanent shops at the sides and rear of the square. If he wants ironware, he will go to the blacksmith's, whose fireside is a favourite meeting place for friends on a cold morning. In the summer, such 'get-togethers' will probably be at the barber's or shoemaker's. Other businesses carried out in the colonnade of shops will include the carpenter's workplace, the doctor's surgery and the teacher's school.

The master might attend a meeting of the Assembly, act as a juryman in a court case, or merely stroll round, greeting old friends, hiring a servant, or watching one of the professional entertainers – a juggler, conjuror or acrobat.

He might go home for a light lunch or buy himself a snack at a nearby stall. When he has had enough of the agora he will go home or to the house of a friend for the day's main meal and some serious talking and drinking.

Town house

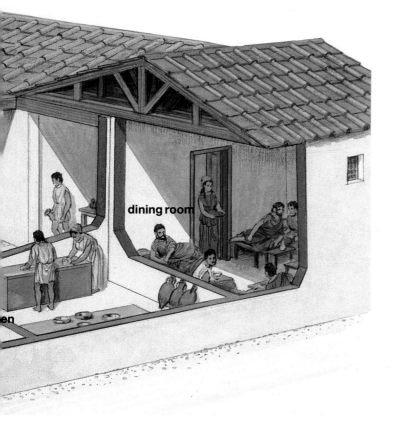

dining room

en

41

Other members of the family

While the master was out at the agora, what were the rest of the family and relations doing? It has to be said straight away that ancient Athens was a man's world: decisions of all kinds, from simple hearth and home ones to those dealing with peace and war were an all-male affair.

A young Athenian woman was married to a suitor chosen by her father. Love hardly entered the matter at all – father picked a likely candidate on the ground of his birth, breeding and worldly goods. Even an old or ugly husband would be acceptable if the other conditions were met.

The wife was expected to raise the children, to feed and clothe the family and to know her place. As a child, she had to pick up what education she could. Girls were rarely educated formally: only her brothers were sent to school, while she stayed at home learning how to prepare and cook food, and how to tailor clothes from the cloth she had woven on her loom.

This is not to say that women did nothing but bathe, chat to friends, paint their faces with cosmetics and have their hair done. Some Greek women were strong personalities in their own right. Women did not vote or speak in the Assembly but there is no doubt that many of them got their menfolk to put forward a particular point of view. There is evidence that women attended the theatre (although never as actresses) and they had their own athletics meetings and festivals.

A Greek woman's sons were only entirely hers until they were about six or seven, when they were sent off to one of the numerous day schools in Athens. Let's ask one young man about his day at school.

'My name is Timon,' he says, 'I am eleven years old and I have been at school now for nearly five years.'

'What time do you get up?'

'I rise at dawn. Our rooster wakes me up early. My father says we borrowed rooster breeding from the Persians, but how my ancestors got themselves up before we had chickens, I don't know.

'I wash in cold water and my slave pours water all over me when I've finished. I dry myself and put my tunic on. In winter, I wear a cloak as well – and sandals also. In summer I go barefoot indoors and out.

A woman weaving

'I eat a bread roll and wash it down with wine for breakfast. I say "wine" but it's really only water coloured pink with a little dash of wine. Even the grown-ups don't drink wine neat: they dilute it with more than the same amount of water.

'I walk to school which is a small corner of a building in an arcade of shops just off the agora. I give my personal slave, or pedagogue, as I call him, my writing tablets to carry on the way. He will protect me to and from classes and wait while I learn. I give the teacher the coins he charges for his lessons and we begin.'

'What do you learn?'

'History and Greek literature – and of course, how to read and write. We copy lines of Homer on to these waxed tablet boards with a pointed stick. The other end is blunt and flat. We use it to smooth out the wax if we make a mistake. We learn arithmetic with a bead frame, pebbles on the ground, or even on our fingers. Some masters teach geometry and nearly all of them will tutor us in singing, speaking and playing the lyre or flute.

'Later on we strip off in the nearby gymnasium and practise running, jumping, wrestling, boxing, throwing both the javelin and the discus and many other sports.

'Then I wash, dress and go home. I can play toy soldiers with my brothers, while my sister (provided that she isn't helping my mother) has some rather nice dolls to dress and play with.

'If there are no guests, we children sit up to the table on stools, while father reclines on his couch. If there are guests, we eat separately before going to bed for the night.'

A Greek school

43

Section 4 *Greek civilisation*

Art and pottery

When we looked into an Athenian house (see p. 41) we mentioned that most families had a collection of rather fine decorated vases. The potters of Athens were so famous for making these that their output was sent all over the Greek world by traders and merchants.

Although the earthenware is only one example of Athenian art, which also includes painting, coinage and sculpture, it is most important for us because it is the the one thing that has lasted down the centuries. A good deal of our knowledge of Greek everyday life comes from the pots painted with human figures, usually black on an orange background or the other way round.

Woman spinning

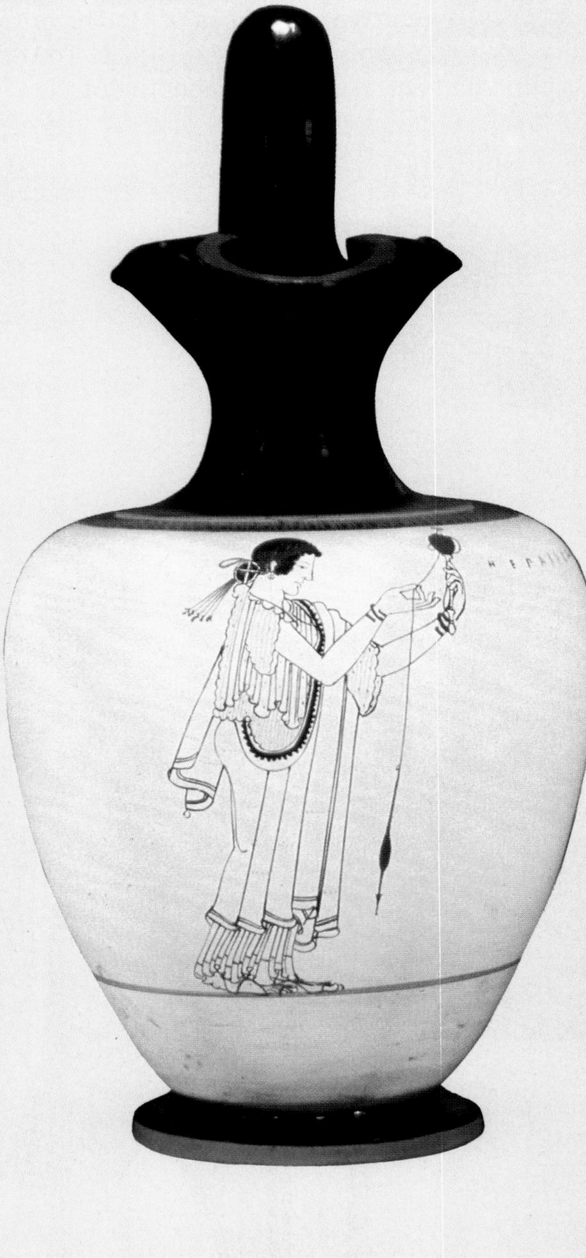

Cobbler

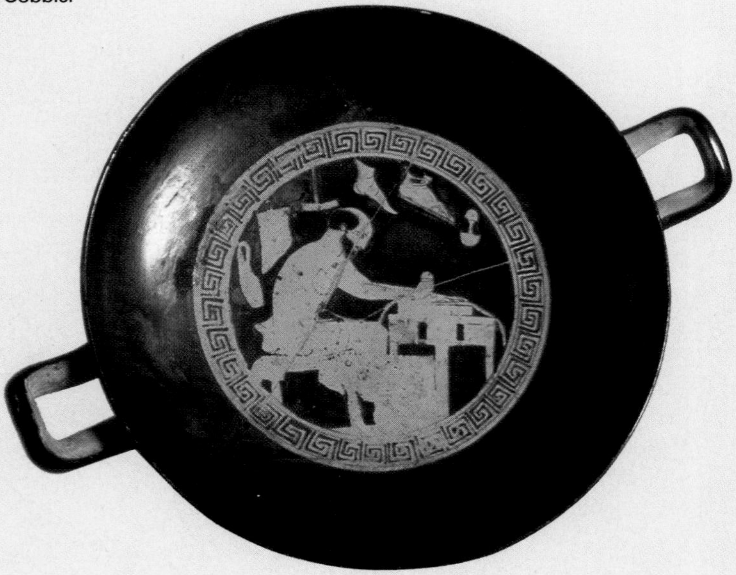

Women fetching water

Fisherman

Huntsman

Carpenter

Ships and trade

As the population of Greece grew, the country was unable to produce enough food for all the extra mouths. Trees were cut down for charcoal (used for metalworking and cooking) and timber (used for house and ship building): much of the soil was washed away by winter rains.

In fact, the only crops that would grow at all well were grape vines for wine and olive trees for olive oil – the cooking, lighting and washing oil of the ancient world. It was a matter of common sense to send these plentiful products abroad and to import the grain which Greece needed.

Exports included not only olive oil and wine but also silver from the mines at Laurium near Athens. One other export has already been mentioned – the huge quantities of painted pottery which found its way all over the Greek world. Athens took over the role of chief pottery maker from the city of Corinth in the middle of the 6th century B.C. and kept it for more than two hundred years.

By now Greece had colonies all round the Mediterranean. We've already seen how adventurers settled on the west and south-west coasts of what is now Turkey, together with the islands of the Aegean. Colonists also took over the islands off the opposite, or western coast of the Greek mainland, settled in southern Italy, North Africa, Spain and cities such as Massilia (Marseilles) in the South of France. Stranger still, perhaps, to modern thinking, they started building trading posts and versions of the Greek 'polis' around the shores of the Black Sea in what is now Russia.

The Black Sea colonists were following in the footsteps of the mythical Jason and the Argonauts who sought the golden fleece at a place called Colchis. The voyage was repeated in the 1980s by Tim Severin and his volunteer crew sailing a specially built reproduction of an ancient Greek merchant ship. The modern voyagers didn't meet multi-headed serpents nor fight warriors who grew from dragon's teeth but they did prove the journey, though difficult, could be done (see

Greek merchant ship at sea

p. 68). In fact, Athens got a good deal of the grain its citizens needed from south Russia.

Greek merchant ships were less than one hundred feet long, with a rectangular sail on a single mast. The sail was mostly used in favourable winds, the men rowing only when the wind was against them or when manoeuvring.

Sea travel was uncomfortable and dangerous. Ships had open holds and benches for the rowers. There was little in the way of decking, the crew sheltering under a spread canvas if it rained hard.

At night, or in really bad weather, the vessel was beached and hauled up on the sand. This was because Greek captains tried always to keep in sight of land when sailing, which isn't possible in the dark. Of course there were no compasses or other navigational aids. They managed perhaps fifty or sixty miles a day, on average.

Athens imported metals and metal goods, hides, furs and slaves, as well as grain. The last mentioned accounted for two thirds of all the cereals Athenians needed – some from Egypt, some from Russia. Not long ago, archaeologists discovered the ruins of a Greek grain port at Olbia in south Russia.

Greek shipwrecks of the period are occasionally found. One such vessel had carried wine in large, clay jars called amphorae. A few were unbroken and still stoppered with the wine inside. Needless to say it hadn't 'kept' and was quite undrinkable.

Military ships were a little longer than cargo vessels and had part of the prow sticking forward some ten or twelve feet. The commander tried to use this 'beak' to ram an enemy ship. Failing this, he aimed to row alongside it, ordering his men to 'ship' their oars at the last possible moment. With his own oars out of the water, and by skilfully using the steering sweeps, he hoped to snap off the enemy's oars like carrots.

All Athenians had to do military service which might be in the army or the navy. However the rowers were paid professionals: they had to be when vessels with two or three banks of oars were introduced (biremes and triremes). A citizen could find himself serving as a sailor one day and elected admiral the next. Of course, he could be just as quickly reduced to the ranks again, especially if he were unsuccessful.

Divers using water probes to investigate Greek shipwreck

The Gods

Hera and Zeus

Athene

Apollo

The very earliest Greeks probably worshipped an earth mother goddess but later invaders brought a male deity with them. In the classical period, religion seems to have descended from a mixture of the two beliefs.

Let's ask this Athenian what he believes in and worships. 'Is there a service in your temple today?' we ask him.

'Service?' he says, 'I don't know what you mean.' We explain and his face clears. 'Oh no,' he says, 'The temple is the home of the god and shows how important the town is but we don't worship in there.'

'Where do you worship then?'

'At any shrine or altar of the particular god we want to talk to. I'm not sure that 'worship' is the right word: we may want to thank the god, to praise him, or plead for something – maybe even bribe him.'

'You have lots of gods then?'

'Oh yes. Some say there are almost as many of them as there are people in Greece. Of course, there are important ones who are known all over the land but every town, village, house, field, fountain, stream, forest, or what you will, has its own spirit.'

'Tell us about the well-known ones.'

'Well, to start with, there's Zeus, the father of the gods. He lives on Mount Olympus with his wife Hera and about ten other Olympians, as we call them. Zeus is the king of the gods and controls the weather, sending sunshine or thunder and lightning at his will. Hera ill treats the girls her husband falls in love with but she also looks after married women and is interested in the bringing up of children.'

'Who are the other deities?'

'There's Apollo, the sun god, who is supposed to be the patron of archery, music, truth, fortune telling, healing, law and order and moderation in all things. His twin sister is Artemis, the goddess of hunting, guardian of cities and young animals, protector of women.'

'How can she guard animals if she's the goddess of hunting?'

'You have to forget one of those if you're interested in the other one – just close your mind. Now, where was I? Oh, I know – Aphrodite, goddess of love and beauty, Demeter goddess of farm crops. Would you like to hear a story about her?'

We nod and he continues. 'Hades, lord of the underworld which was named after him, carried off

Hades carries off Persephone

Demeter's daughter, Persephone, to be his bride. Demeter then refused to make the corn grow or the fruit to ripen until she got her daughter back. The people of the land of Greece began to starve, so Zeus talked to Hades and got him to agree to let Persephone come back to the world for eight months of every year. The remaining third of the year she was to spend in the underworld. This, we believe, is how the seasons started.'

Privately, we think this might be a folk memory of ancient famines caused by climate changes or soil erosion but we let him go on.

'Hermes is the son of Zeus and the messenger of the gods. He also protects flocks and herds and is the god of trade and the market place. Among those who look upon him as their champion are mischief-makers, wayfarers, orators, thieves and writers!'

'Poseidon is the god of the sea, Ares the god of war. Hera, his mother, hated Ares for the deaths he caused but Hades loved him for the extra souls he sent to the underworld! Hephaestus, the lame, is the god of fire and of blacksmiths and armourers. He is a clever craftsman and is supposed to have made Pandora, the first mortal woman. Dionysius, god of wine, is a symbol of revelry who gave man a gift which could be used or misused.

'There are many other supreme beings who are widely known – for instance, Pallas Athene, the patroness of our own city of Athens. Then there's Prometheus, the original fire god, who was dismissed and punished for giving Mankind the secret of fire. Iris is the goddess of the rainbow and Pan, the god of flocks and woodlands. The lesser spirits include Naiads and Dryads who were nymphs of water and trees. I suppose you've heard of centaurs? You know – half man and half horse?'

We nod but we think centaurs may be another folk memory – perhaps of the first Greeks ever to see a man riding a horse. Our musings are cut short as our friend speaks again.

'Also we sometimes revere the old heroes – Achilles, Heracles and—'

'Hold on. That's a lot to take in.'

'There's more.'

'I'm sure there is but I reckon that's enough to think about for the time being. Thank you for your information.'

Section 5   *The Persian wars*

The Persian empire

A map showing the Persian Empire at its greatest extent

BLACK SEA

ASIA
MINOR

LYDIA

Sardis

Tarsus

Nineveh

GREECE

CRETE

Sidon

Tyre

Baby

MEDITERRANEAN SEA

AFRICA

R. Nile

RED SEA

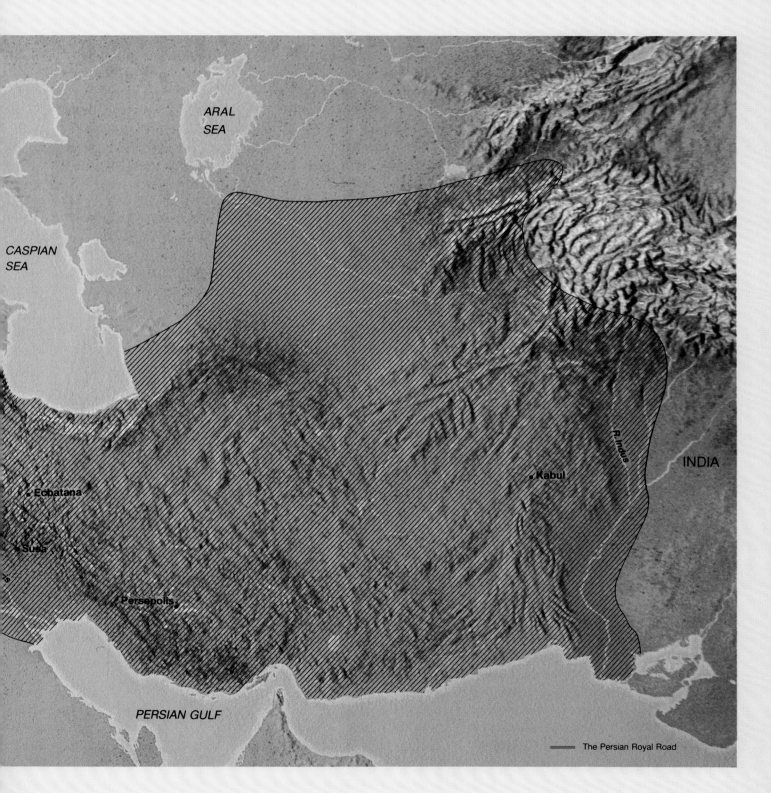

ARAL
SEA

CASPIAN
SEA

INDIA

R. Indus

Kabul

Ecbatana

Susa

Persepolis

PERSIAN GULF

—— The Persian Royal Road

Section 5 *The Persian wars*

Persian rulers

Cyrus the Great

Persian soldier

It seems incredible that an unimportant tribe in the near east should have been able to conquer its neighbours near and far and to become the largest and most powerful empire the world had yet seen. Even more remarkable is the shortness of the period during which Persia was the supreme country.

Its greatness started with the conquests of its mightiest monarch, King Cyrus, in the middle of the sixth century before Christ. This magnificent empire was conquered and finally shattered, lasting only a little over two centuries.

The period is so short that a list of a mere dozen or so rulers is enough to cover it. The first king (as we've seen) was Cyrus the Great, who set the empire on its way by beating the Medes who had been the supreme power in the area. The last Persian monarch was Darius III, who was killed in 330 B.C.

The reason why the Persians were able to win battles so easily was to be found in the way they fought them. Every Persian had to be ready to fight – rich men had to provide their own horses and horse trappings. These became the cavalry, whilst ordinary peasants fought with short daggers and lances. Everyone had to learn archery, for bowmen were often the key soldiers of the Persian army.

Although many of the country's rulers were extremely good generals, it was the tactic which made use of archery that won battles. The cavalry might try to surround or drive off enemy mounted troops but first, the infantry wedged their wooden, hide-covered shields in the ground in front of them to act as a barrier. Then they loosed swarms of arrows at their foes.

In most cases, they never had to get into a situation where they had to engage in hand-to-hand fighting, because their enemies (mostly armed with long spears) never got near enough to use them.

Cyrus, who founded the Persian empire, did it by also conquering Babylon, Afghanistan and the Greek colonies in what is now Turkey. His son, Cambyses, defeated the Egyptian pharaoh, Psamtik III but failed to take Ethiopia. His army was destroyed by a sandstorm in Libya and he, himself, died in Syria on his way to crush a revolt.

After his death, a usurper tried to seize the throne but Darius I, a cousin of Cambyses, beat the rebels and restored the family to the throne. He divided the country into twenty districts called satrapies. These satrapies were occupied by the peoples originally subdued by the Persians. They paid taxes whilst the Persians did not.

Darius reorganised laws, postal services, weights and measures and set up a proper system of money. Although some of his foreign troops accepted gifts of things such as food, land, houses, furniture and so on as wages, others (in particular, his Greek subjects) insisted on payment in gold or silver coins.

Both Darius and his son, Xerxes, tried to conquer Greece, neither with final and lasting success.

Many of the remaining monarchs were named in honour of these two men until the last of them, Darius III, was defeated by Alexander the Great.

This impression from an engraved chalcedony gemstone shows a Persian horseman attacking a Greek soldier

The Battle of Marathon

The Greek colonies in Persian Asia Minor felt themselves threatened by Darius I. They were told that they must not only pay tribute to the Persian empire but also do compulsory service in the Persian army. In the year 499 B.C. they revolted against their rulers.

Athenians and some other Greeks from the mainland sent help and together they advanced on nearby Persian towns and sacked them. Unfortunately, the mainland Greeks then decided to leave their kinsmen to carry on the struggle on their own and they sailed back home.

Darius's army made short work of the rebels, destroying their fleet and capturing and burning Miletus, one of the chief Ionian towns. Many of the beaten Greek colonists were then exiled to the mouth of the River Tigris – over a thousand miles away.

When the news reached Darius in his capital city of Susa, he gave orders for the conquest of Greece. He sent his son-in-law, Mardonius, to start the invasion. Mardonius overran Thrace and Macedonia in the north of Greece, but then had his ships wrecked and had to give up any idea of going further south into Greece.

This was in 492 B.C., and shortly afterwards Darius sent ambassadors to the remaining Greek city states, demanding that they recognise him as their overlord. The messengers were told to ask for presents of Greek earth and water as a sign that the Greeks agreed to Persian rule. The envoys were siezed by the angry Greeks and thrown into deep pits and wells, with the message: 'If you want Greek earth and water, help yourselves!'

Persian envoy being thrown into a deep pit.

Darius was hardly pleased at this turn of events and in 490 B.C. sent a huge army in 600 ships under the command of his generals, Datis and Artaphernes. The plan was to land at the bay of Marathon and march overland to Athens.

The Athenians awoke to their danger almost too late. Miltiades, the Greek general, who came from Thrace in the north, had some experience of Persian battle tactics. He thought the best thing was to hit the Persians rather than wait for them to attack first.

While working out his plans, he sent Pheidippides, a well-known athlete, to run to Sparta seeking help. Pheidippides ran almost continuously for two days and nights only to be told by the Spartans that they were having a religious festival and that they couldn't start out until the next full moon. By the time Pheidippides had run back to Marathon he had covered over a hundred miles. He was also to fight in the battle.

Miltiades made an attempt to cut down the Persian advantage in numbers by stretching his lines of men across a narrow valley. He put his weakest formation in the centre and his strongest ones on the two wings.

The Persians were delighted to see the weakness of the lines in front of them and made ready to attack the Greek centre. However, the Greeks began to run forward when they were still just out of range of the enemy arrows. Their centre was swamped but the two wings closed on the Persians like the jaws of a vice. Vicious hand-to-hand fighting broke out and at the end of the battle, the Persian survivors lost their formations and ran for their ships, leaving 6,400 of their comrades dead on the plain of Marathon.

The Athenians lost only 192 men. They were buried in a common grave and the earth heaped up over them. The burial mound is still in existence and can be seen by the visitor to modern Greece.

The great empire of Persia was beaten but it wasn't a final victory for the Greeks, only a breathing space.

As for poor Pheidippides, he managed his long distance run and even survived the battle. Unfortunately, he was then sent to take the good news of the victory back to the citizens of Athens. He ran all the way, gasped out his tidings and then fell dead.

Thermopylae

'My name is Ninaku. I'm an archer from central Persia. Ten years after our disaster at Marathon, Darius's son, Xerxes, our new king, has decided once and for all to beat the proud Greeks to their knees.

'Those Greeks seem to us simple soldiers to be far too cunning for their own good. I fought at Marathon and I should know. After that battle, we went on board our ships and sailed for Athens, hoping to catch them by surprise but the devils had marched back and were waiting for us. We had to leave empty-handed that time but we won't be put off on this campaign.

'Let me tell you how Xerxes got our huge army over from Asia to Europe. There's a stretch of water just over a mile wide between the continents. It's called the Hellespont. Instead of sailing across, the king gave orders to make two boat bridges. Each one had more than 300 vessels held in place by stem and stern ropes tied to heavy boulders which were then dropped down to the sea bed. The ships were joined side by side with thick ropes across which stout planks were laid.

'These were spread with straw, topped with a thick layer of earth. This was supposed to help the horses, as were the canvas sails rigged up on each side all the way across to screen the sight of the sea from the animals.

'You can imagine the chaos if one or two had reared, shied or bolted! I don't know about the horses but there were a few of us who were happy not to see the waves. I can remember feeling uneasy in one place where there were gaps between the planks. It was bad enough to catch sight of the water below but I couldn't help thinking we were half a mile from the nearest land – and I can't swim.

Crossing the Hellespont.

'At last all the army was over – Ethiopians draped in leopard skins, us Persians in our woven cloth uniforms, Indians in cotton materials, other warriors in goat skins – you never saw so many styles of dress, nor so many shades of skin colouring – all the way from dark brown to light pink.

'I've heard a rumour that there were two and a half million in our army – one fellow thought it might be more than *five* million! All wrong, of course. Men came ashore on the Greek side at the rate of about one every three seconds: admittedly, they were doing that non-stop for a week but even so, it only comes to about two hundred thousand men. Work it out for yourself.

'We moved southward down the coast till we came to a narrow pass between the mountains and the sea. Our spies told us that the Greeks call it 'Thermopylae', which means "the gate by the hot springs".

'There we found three hundred Spartan warriors under their king, Leonidas. We tried to force our way through but the pass was so narrow, we couldn't use our superior strength.

'Then we had a stroke of luck. After several unsuccessful attempts to smash the Spartans we sent a spy to look over their camp. He reported back that they were doing exercises and combing their hair. Xerxes laughed until one of our captains told him that the Spartans only did that when they were prepared to fight to the death. Xerxes grew thoughtful but then our spy produced a Greek traitor he had met who was willing to show us another way through the mountains.

'We moved in single file along the steep rocky paths until several thousand of us had passed. Then we attacked the Greeks from front and rear at the same time. They fought like tigers until their weapons were broken and useless. Then they used their bare hands.

'At last they had all fallen and the way was clear for us to descend on Athens. We were sure it would be easy this time, as the rest of the Greek army had withdrawn to defensive positions well to the south of Athens. Nothing would stop us now.'

Section 5 *The Persian wars*

Salamis

Women and children being evacuated

Realising at last that nothing could now keep the Persians from their city, the Athenians put a bold plan into execution. The first part was to evacuate the women and children to nearby islands and enlist nearly all the able-bodied younger men into the navy, which was still in existence and so far unbeaten.

Only a handful of defenders guarded the deserted streets of Athens. The Persians swarmed in through the gates and attacked the Acropolis where remnants of the Greek garrison were holding out. So well did they fight that it was a fortnight before they were overwhelmed.

The Persians then proceeded to smash, burn and destroy every monument, building, temple and statue they could find. Destruction was nearly complete. When the Greeks finally retook their city, they allowed some of the damage to remain unrepaired for decades as a warning to the citizens.

Themistocles, who had risen to power during these troubles, resisted demands that the Greek fleet be sent to support the remainder of the united cities' army, now guarding the Gulf of Corinth. He wanted to engage the Persian ships at a place called Salamis, not far from Athens.

It was a narrow strait between the island and the mainland and Themistocles reasoned that, as at Thermopylae and Marathon, narrowness might prevent the Persians deploying their full strength.

His problem was to trick the Persians into attacking him at the place he had chosen himself. He solved it by sending a messenger to Xerxes. The messenger pretended to be a traitor. He told the Persian king that he was in sympathy with the invaders and that the Greek sailors were so terrified of the Persians that they would sail away at the first sign of trouble.

Xerxes gave orders straight away to set sail for Salamis and to attack the Greeks immediately they arrived. The second part of the cunning plan of Themistocles was to tell his 'traitor' to let Xerxes know that if a Greek fleet were to be attacked without further ado, a good half of the sailors would change sides.

The Persians sailed confidently into the narrows at Salamis to meet their enemy. Greek ships appeared in front of them and then suddenly, another contingent showed up behind the invaders. Boxed in, they were at the mercy of the Greeks. The Athenian ships had rams and were smaller and more manoeuvrable than the clumsy Persian galleys.

Xerxes, watching the battle from a throne on the shore, could hardly believe his eyes. His fleet lost at least half its ships, the Greeks scarcely forty.

Xerxes, in despair, left for Asia, taking most of his fighting men with him. He left some under the command of Mardonius to renew the attack the following year. Facing the Persians was the combined Greek army of almost a hundred thousand men, led by Pausanias, a Spartan general. They were still outnumbered by their enemy but not by nearly so many as they had been the previous year.

The two forces met at Plataea. Mardonius decided to wait until the Greeks attacked and then destroy them with his mounted troops. It was a good plan but it came to nothing because of a Persian error. When the Greeks began to rearrange their battle order, Mardonius mistook the troop movements for a Greek retreat. He forgot all about waiting and gave the order himself to start the battle.

It was the wrong move. Pausanias's men took the shock and then counter attacked. Mardonius was killed, together with most of his foot soldiers. The rest fled.

The Greeks were not to know it but this battle in 479 B.C. was the last main engagement of the wars. From that time on, Persia was never again a serious threat.

Section through trireme showing rowing positions

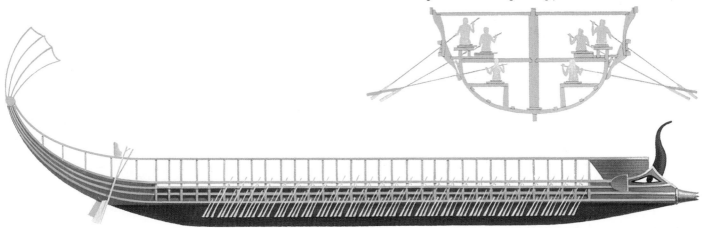

Athenian ship showing the ram at the prow

The Persian ships suffering defeat in the straits of Salamis.

The wise men of Greece

Following the defeat of the Persians, there were almost fifty years of peace. Athens became the leading city state and enjoyed a period of prosperity together with a flowering of the arts and sciences such as the world had never seen before.

Under the leadership of Pericles, the city was beautified, its architects designed and erected splendid public buildings, and sculptors created many marble and bronze statues to adorn them. Its thinkers put forward all kinds of new ideas and its dramatists presented plays both serious and comical, so excellent that many of them are still performed today.

Not all the outstanding people listed below belong to this Golden Age but the fact that such a high proportion of them do, shows what a rich half century it was. The actual dates are 479 B.C., when the Persian menace came to an end, and 431 B.C., when the Peloponnesian war broke out. Perhaps you can work out which of these lived and worked during this period?
(Note: all dates given are B.C.)

Kallikrates (5th century) A master builder and architect whose designs for the new Parthenon and the temple of Athena Nike on the Acropolis were accepted and used as part of Pericles's plan to make Athens the most glorious city in the world.

Ictinus (5th century) Many considered him the finest architect of his day and he was entrusted by Pericles with the construction of the Parthenon. He also helped design the Parthenon.

Myron (5th century) A sculptor whose bronze statues of animals and athletes were always in demand. His best known work, copies of which have survived, is the 'Discobolus' (discus thrower).

Pheidias (c490–432) This sculptor was a native Athenian. He worked on the statue of Athena Parthenos. When the architects had put up their buildings, he was entrusted with all the artistic decoration – including the friezes on the buildings of the Acropolis.

Herodotus (c484–424) Known as the 'Father of History', his work on the recent past of his own countrymen is one of the chief sources of information for what happened in the Persian wars.

Pheidias carving a statue

Sophocles

Sophocles (495–405) Born near Athens, he became a great dramatist. He is supposed to have written over 130 plays, of which only seven have come down to us. Among the most famous of them are: *Antigone*, *Electra* and *Oedipus Rex*.

Anaxagoras (c500–428) A philosopher (Greek word meaning 'lover of knowledge') his idea was that the universe was made up of little beads of different kinds of matter. These were all mixed up to start with but something began to move them about and all the similar beads drifted together to make the things we see. He was condemned as an atheist ('anti-god') when he said that the sun and the moon were not divine persons but fiery, molten lumps.

Thales (640–550) A philosopher who was the first to suggest a scientific explanation of the universe rather than one of myth or religion.

Socrates drinking hemlock

Socrates (c470–399) A philosopher who conducted debates with his young followers on morals and religion. He was found guilty of corrupting youth and sentenced to death. He chose to kill himself by drinking hemlock.

Plato (427–347) Travelled widely and at one time was sold as a slave. He became a disciple of Socrates, whose work he carried on. His interests included the things that affect our conduct and character. He believed that the good citizen could only flourish in a just and orderly society.

Pythagoras (582–?) He had some strange ideas about mathematics, dieting, religion and music. His famous theorem dealt with the relationships of the sides of a right-angled triangle.

Aristophanes (c445–385) He wrote 54 comedies for the stage, of which only eleven survive. The best known are *The Birds*, *The Wasps* and *The Frogs*.

Euripides (480–406) 18 of his 90 plays still exist and are still performed, including *Medea*, *The Trojan Women*, *Orestes* and *Bacchae*.

Aeschylus (525–456) Author of more than 70 tragedies. We know of only seven, including *The Persians* and *Prometheus Bound*.

Aristotle (384–322) Probably the first man to advocate the scientific method: 'Look carefully first, then make your theory'. Tutor to Alexander.

The doctor and patients

Hippocrates (c460–377) 'Father of Medicine', he was both physician and surgeon. We know of 70 of his essays on medical subjects.

Democritus (c460–357) The first to say that everything was made of atoms.

Thucydides (c464–c404) Wrote a fair and accurate history of the Peloponnesian war.

Empedocles (c493–c433) A philosopher who held that everything is made of earth, air, fire and water.

Demosthenes (383–322) Orator and statesman. An orphan, he studied law and to cure his stammer, he walked the seashore with his mouth full of pebbles and tried to outshout the noise of the waves crashing on to sand and rocks.

Slavery

Slaves working in the mines

Slaves working on a farm

My name is Enkales (Enk-a-lees) and I am a slave. I'm a slave because my father was a slave and his father before him. It's said that my grandfather came from somewhere to the south of Egypt once long ago. He was on a voyage across the Mediterranean when his ship was attacked by pirates. They stole the cargo and sold all the people into slavery.'

'Is that how people become slaves in Greece?'

'Not the only way, no. You could be kidnapped, captured as a prisoner of war or sold into bondage for debt. I don't think the last one is still used and I was told that the great Solon had done away with the custom of making a slave of any labourer who didn't work hard enough.

'It was Solon who freed all kinds of slaves; for instance, any Athenian who had to flee abroad to avoid slavery, any Athenian who had been sold in a foreign slave market or any Athenian who'd been made a slave here in Attica.

'My job is to look after the house, help with the shopping, chopping wood and getting it in, buying charcoal for the cooking fire and fetching all the household water from the public fountain. I also do all the odd jobs that crop up around the house as well as walking the young master to and from school when his regular slave can't do it.

'It's hard work most of the time but I suppose I'm better off than some slaves. I'm talking about those who work in the mines. I overheard a house guest the other day. He had rented part of the silver mines at Laurium. He and scores of other mine men bought up to a dozen or so slaves each and set them to work. They had absolutely wretched conditions to toil in.

'Most of the time they worked until they dropped in hot, wet darkness, breaking their backs to fill leather bags with lumps of rock which they had smashed with pickaxes and crowbars. Their owners' attitude seemed to be "Slaves are cheap, so keep them at it until they fall exhausted." In fact, slaves aren't all that cheap—'

We interrupt to try and find out how dear they were

Slave auction

in terms we can understand. It seems that it would have taken the complete earnings of a modern average workman for six or seven weeks to buy a male slave.

'Anyway,' says Enkales, 'there are even now as many as twenty or thirty thousand poor devils slaving away in those silver mines.'

'Couldn't they complain?'

'You obviously don't know much about the set up in Athens. No slave is allowed to vote or speak in the Assembly and as for having a revolt – well! We'd never get enough of them to do the same thing at the same time. Why, they come from so many different parts of the world, you can't even find a language they all understand. Some of them don't speak a word of Greek – especially the mineworkers.'

'Do slaves work on the land?'

'Yes, but not as many as Sparta, or some of the other cities, use to farm with. Even so, it's been worked out that almost half of all the people who live in Athens are slaves. Many of us are house slaves and others work in quarries, docks or workshops, making things.

'I suppose the owners of these businesses have to be careful not to employ too many slaves or no free man would ever find work. As it is, most households have about three slaves on average.

'A slave can, if he's lucky, buy himself out. All he needs is to lay his hands on a lot of money. I've already said that house slaves are better off than those who work in quarries or mines but all of us suffer in a way I've not told you about yet.'

'What's that?'

'Well, if ever any one of us is called upon to give evidence in a court case——are you sure you don't know about this?'

We shake our heads.

'The rule is, that however a slave is connected with a trial, even if he's only an innocent bystander, he must be tortured first before his evidence is taken, because they believe he will tell lies otherwise!

'You can see that it's not such a golden age for us!'

Clothes

Clothing fashions changed very little throughout this period – certainly not as much as they have in modern times. Greek clothing needed no skilled tailors to make it fit and yet it remains among the most elegant of styles that human beings have ever invented. In fact, sculptors and painters of the last few centuries of our own age have sought to give their, often royal, subjects grace and dignity by showing them in Roman toga and tunic, or in the Greek equivalent of himation and chiton.

A chiton (pronounced kye-ton) was a loose tunic worn next to the skin by men, women and children. It was not much more than a tube of woven material of either linen or wool, with armholes or short sleeves. It was pinned or sewn together over the tops of the shoulders. For men and boys it did not reach the knee and was either its natural light colour or bleached white. For women and girls it was ankle, or even floor length and might be dyed or patterned in colours.

In both cases it was hitched up with a simple girdle of leather or cord around the waist. No underclothing was worn – if the person concerned was cold in the winter, he or she might wear more than one chiton.

Over the chiton went the himation, a woollen garment that was merely draped on the body, round the waist, over a shoulder, under the opposite arm and then back over the same shoulder. It was recognised as a sign of elegance and good breeding to have your himation arranged properly: too short and your neighbours sniggered; too long and it dragged in the dirt.

In the latter case, it meant a visit from the travelling laundry-man, or fuller, who collected the soiled clothes, and treated them with various substances such as nitre, potash, or an aluminium salt, known as fuller's earth. With the latter, the fuller made a paste which he dabbed or spread on the dirty linen, particularly if there was a greasy stain to be removed. When the stuff was dry, it was broken and crumbled, the last traces being removed with a wire toothed comb or brush. Then the whole garment was rinsed in clear water. If a bleach was desired, the cloth might be hung up in sulphur fumes.

These processes, unfortunately, left the materials rather less waterproof than when they were first made, so that wearers were more anxious to find shelter when

Greek dress for women

Fullers at work

it rained. Strangely enough, long sleeves, which would have given some protection, were never fashionable, being left to workmen to wear. The chitons workmen wore were usually dark-coloured so that they didn't show the dirt.

Clothing was made by hand, either in a workshop or by the farmer and his wife. If they didn't own sheep, poor peasants probably wore clothes of soft leather.

As sheep had to be sheared, the wool washed, combed, spun and woven, the resulting garments were extremely hard wearing but rather expensive.

Because of their cost, the Athenian's himation and chiton were a tempting target for thieves. If a citizen bathed or took exercise, he did so naked and should have had a slave to guard his things. In the absence of a guard, the clothes were liable to be stolen. In fact, some thieves were bold enough to snatch the himation from a man's back in the street, particularly at night.

In the house of a friend or at home, the Athenian went barefoot but a sort of sandal was worn in the street. Country folk wore stouter boots and poorer peasants wooden-soled 'flippers' held on with thin leather thongs. If a cobbler was making a pair of shoes, he would ensure a good fit by asking the customer to stand on the leather. He would then cut round the shape of the customer's foot.

Beards and hair had been long at the time of the Trojan war but they were shorter during the Golden Age. Men didn't often go clean shaven until the age of Alexander. Women tried a variety of hairstyles and were fond of wearing jewellery, especially engraved precious and semi-precious stones.

By the way, if you saw an Athenian with very short hair, he might be in mourning or he might be a mean man trying to save money on trips to the barber.

Greek dress for men

Greek hairstyles and beards

The Acropolis and its buildings

We've heard how the Persians had left Athens in ruins. The defenders of the inner fortress, or Acropolis ('highest city'), had held out for several days before finally giving in. It seemed that the temples and other buildings on the rock had taken the brunt of the Persians' anger.

For a while, the fire-blackened remains were left as a monument for the citizens to see and it wasn't for decades that a decision was taken to rebuild. In the meantime, one of the destroyed temples was replaced with a temporary wooden building while Pericles looked about for men to do the reconstruction.

The men he chose were Ictinus and Kallikrates to design the new temple and supervise the workmen, plus Pheidias to provide the 500 or so sculptures that were to adorn the finished structure. The work began in 447 B.C. and took fifteen years.

The top of the rock was smoothed and levelled to take the base, then the fluted pillars were put up – seventeen along each side and eight across the front and back. The original measurements of the building were: nearly 230 feet long, about 100 feet wide and 65 feet high. The tops of the 58 main columns were joined by flat stone slabs and a shallow triangle of stone was set up at each end, later to be filled with sculptures. The rest of the roof was made of wood with tiles on top.

The citizen visiting the temple would perhaps make a sacrifice on the altar just outside the main doors before going in to pray to the goddess, Athena Parthenos. No ceremonies were conducted inside the temples.

Inside the rows of columns was a passageway all round the main building. In a back room was a store for the treasures, but in the main room, facing the front door, stood a huge statue of the goddess, at least 40 feet high.

Pheidias had made a wooden framework and covered it with carved ivory for the lady's hands, arms, neck, head and face, plus moulded gold for her head-dress and draperies.

Unfortunately, the statue no longer exists: it was taken away to Constantinople a thousand years later, where it was destroyed by fire some time between the 6th and 10th centuries of our own times.

The ancient Athenian worshipper would have seen the giant statue only dimly. There was no lighting save that which entered through the main doors, or was reflected up from a pool of still water at Athena's feet.

From the outside, the temple still appears in modern times to be a marvel of proportion but the builders had worked cunningly for effects. The straight lines you can see are nearly all a little curved and the gracefully tapering vertical columns bulge at one point and actually lean inward slightly.

With the right weather conditions, the visitor today will experience a stiffish breeze as he goes through the Propylaea (front gates), a magnificent entrance way to the top of the Acropolis. Looking back the way he had come, he would be gazing in the direction of Sparta and somewhat to the left of that, down to the Piraeus, the road to which was once protected on each side by walls four miles long, so that ancient Athens couldn't be cut off from its port and be starved out in a siege.

Passing on to the rock proper, he would see on his right a small temple dedicated to Athena Nike (goddess of victory). To his left was the Erechtheion, with one of its porches supported by Caryatids (carved stone human figures) rather than columns.

Nowadays, most of the sculpture has long since gone. What still exists is in museums and is a plain yellowish white. It's difficult to imagine Greek statues as they nearly all once were – painted pink for skin, yellow for hair and with red and green clothing on a bright blue background.

The Parthenon still looks wonderful but there is nothing inside it. In fact, it's surprising that there is as much left as there is. It remained a Greek temple for a thousand years, was turned into a Christian church for the next thousand and into a mosque during the Turkish occupation. The Turks later used it as a gunpowder store. Then a 'lucky' shot from a Venetian cannon landed right on the explosives and blew the inside to pieces.

The sculptures from the walls of the Parthenon treasury were brought to Britain by Lord Elgin and may be seen in the British Museum. They show a procession of 192 youthful horsemen – probably representing the 192 soldiers killed at Marathon.

Greek legends

Greek parents often told their children stories of Greece's heroic past. The tales were literally about heroes and heroines, although many of these fantasies also dealt with gods and their adventures. The themes of these stories were used extensively by poets and playwrights. Here are brief outlines of two 'hero' stories:

Herakles (called Hercules by the Romans)

His ancestors were descended from Zeus and he was supposed to be the bravest and strongest man who ever lived. He was so hated by the goddess Hera that she sent snakes to attack him in his cradle but the baby Herakles strangled them. Because of a crime committed by his father, he was ordered to make amends by carrying out a series of almost impossible tasks, or 'labours', as they were called. There were twelve of them:

1. He had to kill the 'unbeatable' lion of Nemea. He closed with the beast and strangled it with his bare hands.
2. He slew the Hydra, a monster with nine heads.
3. He was sent to capture the golden-horned stag of Arcadia, which he did after tracking the animal for over a year.
4. Another animal's capture was demanded – that of the giant boar of Erymanthus. He chased the animal so hard it became exhausted and thus easy prey.
5. He was given the chore of cleaning out the Augean stables in Elis, an enormous building with a thousand stalls. He managed this by altering the course of two rivers so that they would run through the animals' quarters and wash them clean.
6. He shot with his bow and arrows the birds that were eating the grain in the countryside of Stymphalus.
7. He captured the bull of Minos.
8. He trapped the man-eating mares of Diomedes in Thrace.
9. He persuaded Hippolyte, queen of the Amazons to give him her girdle.
10. He laid hold of the cattle of the monster, Geryon.
11. He had to obtain the golden apples of the Hesperides. Only one person knew how and where to get them. This was Atlas, whose job was to support the sky on his shoulders. Herakles held the sky for him while he went off to get the fruit.
12. His last labour was to descend into Hell to capture Cerberus, the three-headed dog of Hades.

There were many other stories told of him.

Jason and the Argonauts

There was a mention of Jason on p. 46. Here is a tale told about him and his crew.

Phrixus and his sister, Helle, were escaping from danger on a ram with a golden fleece. The ram could fly. Helle fell off into the sea and drowned. The waters have been called the Hellespont ever since. Phrixus landed safely, made his way to Colchis on the Black Sea and sacrificed the ram. He hung its fleece in a sacred grove where it was guarded by a dragon that never slept.

Years later, Jason was persuaded to go and seek this golden fleece. He had a special ship, the Argo, built to a new design. It was the first Greek war galley. Jason called for fifty volunteers, among whom were Orpheus, Polydeuces and Herakles.

After many adventures, they gained the Black Sea and sailed to Colchis on its south eastern shore. The local king would not give up the fleece until Jason had tamed the royal animals. These were fire-breathing

bulls with bronze hoofs. Jason must harness them and plough the field of Ares. Then he had to sow the field with dragon's teeth.

He did this but was surprised to find a fully armed warrior springing from each sown tooth. However, Jason and his crew slew the warriors and with the help of Medea, the king's daughter, took the fleece and escaped to his ship.

One version has the Argonauts travelling home via the River Nile and overland through Libya to the Mediterranean. Another ending of the story makes them sail up the Danube and down another river to Italy.

One interesting footnote was the discovery that some ancient gold prospectors, instead of 'panning' for the precious grains had actually pegged out fleeces in likely rivers and, if lucky, would take them out glistening with gold. Could the story of Jason be a reflection of this fact, just dimly remembered?

Music

Although some tunes with musical notes have been found at Delphi, we don't really know what Greek music sounded like. There were no record players or tape recorders in those days, so all we can do is to make some guesses based on pictures and writings.

We know that in almost every society without a written language, stories, histories, moral tales and so on were passed on from person to person down the generations. If there was a mistake, a mishearing or misunderstanding, perhaps even a simple failure of the memory, the details could become distorted to a greater or lesser extent.

The best way to learn something by heart is to set it to music in the form of verse. Greeks excelled at epic poetry which was often a straightforward account of some historical happening. Their other verse form was lyrical poetry which dealt more with human feelings.

When poetry was recited to music the accompanying instruments were either a pipe-like flute or oboe, sometimes played singly, sometimes in pairs. The other common instrument was the lyre.

The simpler pipes had holes for the fingers to change the pitch and a musician learned to cover half of a hole to get an 'in-between' note. The average pipe was made of hardwood or bone and had a double reed. It probably sounded a bit like a somewhat nasal 'drone' on a bagpipe.

When a player put two pipes to his lips at once, he obviously got them to harmonise, but which one carried the tune and which the descant or how it was done, we just don't know.

The small portable harp, or lyre, came in two kinds. There was the formal or ceremonial lyre, the 'kithara'. The guitar and zither probably owe both their name and their very existence to this instrument. It had seven strings with a wooden sounding board and a plectrum, for plucking the strings, which was tied to the wood to stop it being lost.

The everyday lyre also had seven strings but it was made of cane, wood and horn with a sounding board made of animal skin stretched over an empty tortoise shell.

Flutes and lyre

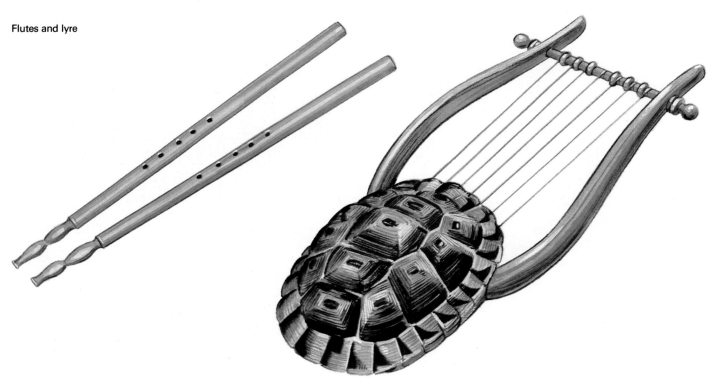

There were probably small drums and the army certainly had a metal horn which could produce two or three notes when the player vibrated his lips and blew harder or softer.

Goatherds and shepherds amused themselves playing on the pan pipes, or syrinx. This was a set of hollow stemmed whistles of gradually increasing length bound together side by side. This is still in use on parts of the Black Sea coast.

The last instrument the Greeks developed was a kind of organ. It was worked by water pressure which drove air into the chosen pipe. The player made the choice by moving a lever.

Going back to the beginning of things, we know that music was important even in Minoan times, for there are pictures of Minoan musicians. By the time of Homer, it's fairly certain that nearly everyone could both play and sing. Music was always a main subject in Greek schools and also a necessary part of everyday life.

Hymns, folk and work songs were sung at various festivals, whether of a religious or sporting nature. They could be sung at weddings or after a successful harvest of grapes or olives. Athletes and warriors often trained to a rhythmic melody. Even drama, which the Greeks practically invented, grew out of music and dance.

Pythagoras, the mathematician, made an important musical discovery. If you stretched a string and arranged it to sound 'C' when plucked, you could divide it into 120 parts and produce a range of notes over a couple of octaves by 'stopping' the string at 30 parts, 40, 45, 60, 80 and 90. He was the first person to discover the octave itself.

It is interesting that the words 'chorus' and 'choir' come from Greek words, as do 'harmony', 'orchestra' and 'music' itself.

Double flute player and lyre player

71

A day at the theatre

Cimon is a mask maker. He is going to do two things: he'll show us one way of making masks for the stage actors and tomorrow he is to take us to the theatre.

We are in his workroom and he is about to start work on the face of the actor sitting on the stool. He begins by rubbing olive oil on to the actor's face. He is a young man and will take the part of a queen. Women do not act in ancient Greece. They are allowed to watch the play but not to take part.

Cimon ties narrow strips of linen around the young man's head until his face is almost covered. Only the nostrils and the mouth have been left free. He then puts on further strips, criss-cross fashion, which have been dipped in a flour and water paste. While he is doing this, he talks to us.

'You'll have to get up early tomorrow,' he says, fixing another strip in place. 'The theatre doesn't have much in the way of lights, so the performance starts at dawn and goes on all day.'

'What a long play!'

'No, it's three plays – four really – three tragedies and a farce at the end. Don't forget to bring some food and a cushion.'

'How did drama start in Greece?'

'Some people will tell you the origin was in Crete but here, in Athens, it began with hymns to Dionysus, the god of wine, for a good grape harvest. Everyone came together to sing and praise the god. Some danced as well.

'Eventually the celebration was held in the agora. A fellow called Thespis started to sing or speak on his own. He stood on a farm cart and did solos.'

'Actors are often known as Thespians, ' mumbles the young actor.

'Keep your jaw still. I'll do the talking. Where was I? Oh, I know – I was saying about talking on your own. It wasn't long before the hymn had turned into two or three solo parts for the actor, with a chorus making comments and descriptions of things the audience couldn't see.

'There were fifty in the chorus and they danced and sang in a circle about eighty feet across, at a place just below the Acropolis. People sat on the hillside looking

Making a mask

down on the "orchestra", as we call the circle of the chorus. In the middle is an altar to Dionysus.

'The next developments were a raised stage for the actors and plays not necessarily about Dionysus, or indeed any god. The audience were given wooden benches to sit on and eventually horse-shoe shaped rows of stone seats were made. A "skene", or small building was put on the higher stage, from where the actors could come on and go off.

'Mostly, we see tragedies – by Aeschylus, Sophocles or Euripides. When there's a festival like the one tomorrow, we put one of our archons in charge. All the authors send him a copy of their new plays and he picks the ones that are to be seen. He also gets in touch with a rich citizen, or choregus, who'll pay for the production. The city provides the actors' wages and even gives ticket money for the very poor. Everybody has to attend.

'Costumes are elaborate, in bright colours and often padded. You may be sitting in the back row and you must be able to see which character is which. You can hear all right, no matter where your seat is.'

'Oh,' we say, 'is that why they have masks?'

'Of course. You may be too far away to see the face, so we make them larger than life – even grotesque. There are only two or three actors, which means each will play more than one part. I've made masks with two faces, so the actor can quickly change expressions or characters by moving the mask round.'

As he speaks, he is cutting through the linen strips at the back of the actor's head. 'Now,' he says, 'I can build up the lips, nose, cheeks, or any other feature with rolls of linen and smooth over all with white plaster. I coat it with a thin glue, let it dry and then paint it. Of course, some masks are made of wood, cork, or even thin metal.' He sets the mask down on a bench and we thank him.

A modern performance of a Greek play in its original setting at Epidauros

The next day we queue up with him before dawn to get into the theatre. He has bought a couple of little, round metal tokens with a seat row and number on them. We take our place, arrange the cushions and slip the food basket behind our legs.

Cimon has a question. 'Your language uses our words for theatre, scene, drama, tragedy, comedy, orchestra, chorus, and many others, doesn't it?'

'Yes,' we nod, 'Why do you ask?'

'Do you know the Greek word for actor? No? Well, it's "hypokrites".'

'Oh yes, of course – one who pretends to be something he isn't!'

Plan of theatre

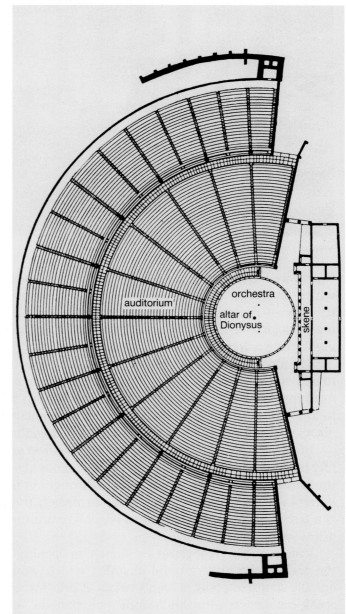

The Oracles

View of Delphi today

The Hypogeum

Greeks were a religious and also a superstitious people, which is strange considering that they also believed strongly in the powers of reason. Because they held many illogical beliefs, they saw nothing odd in consulting fortune tellers to find out about the future.

The fortune teller was usually a priest. He might try to find out what was to happen to you by working out meanings from the insides of animals slaughtered for sacrifice, or from the flight of birds. There were ancient places where many of these priests had come together.

In Greece alone there were over two hundred and fifty temples or shrines where you could find out about the future and there were many others throughout the ancient world. For example, on the island of Malta there is a temple called the Hypogeum, possibly more than 4000 years old.

Visitors to the Hypogeum are shown an opening in the rock of the underground chamber which would amplify sound but only of men's voices. A priest would speak his answer into this natural 'microphone' and the enquirer, unable to see the priest could not fail to be impressed as the reply to his question rolled and thundered around the rocks.

At Dodona, near the modern boundary between Greece and Albania, there was an oracle or seer in the form of an ancient oak tree. Visitors would ask the priests to advise them on a business deal, or about a family or farming matters. The priests would listen to the noise of the wind in the tree's branches and interpret the sounds as a response from the oracle.

The statue of Hermes at Pherai in Thessaly gave advice in an even odder way. The truth seeker put a coin on the god's altar, lit the holy lamps, burned incense and then thrust a finger in each of his own ears whilst hurrying away. At a certain distance he removed his fingers and the first thing he heard anyone say about anything was the answer to his enquiry.

However, the most famous oracle was the one at Delphi, just north of the Gulf of Corinth and set part way up the foothills of Mount Parnassus. There were temples, treasure houses and an athletics stadium farther up the 8000 foot high mountain. In later years the site was neglected and overgrown. A peasant village was built over it. Not until 1893 were the mean little huts removed and the site uncovered. From archaeology and from writings, we have found out a good deal about it.

The place, a fairly lonely and romantic one, was ancient in Mycenaean times. Legend has it that a dragon, the Python, was slain by Apollo who founded a centre there. The chief priestess, the one who made the prophecies, was called the Pythia, or Pythoness.

Private individuals, groups, city councils and even kings consulted her. They wrote down what they wanted to know and their requests were taken into the temple of Apollo, where her underground chamber was to be found. The Pythoness sat on a golden stool and breathed in the vapours that came up from a vent in the floor. To make sure she was drugged, she chewed laurel leaves. Her ravings and mutterings were taken down by the attendants who then said what they thought the meaning was.

The Pythoness

This was then taken out and delivered to the enquirer, often in such a way that more than one meaning could be taken from it. You may remember how king Croesus had asked the oracle if he should go to war against Persia. The Pythoness told him that he would bring a great empire to an end if he did. Croesus took it to mean the destruction of Persia but unfortunately, it was his own empire that fell!

Athenian citizens had enquired about the best defence against Xerxes in the Persian wars. The oracle had advised 'wooden walls'. Fortunately it was decided to build more ships rather than a timber fence all round the city.

The last message that we know of from the oracle was delivered to a messenger from the Roman emperor Julian in 361 A.D. It said, 'Tell your king that the good times are over; that there is no roof nor magic laurel tree for Apollo and that the holy spring has ceased to flow.'

By that time, many previous Roman emperors had robbed the site of much of the treasure given by grateful pilgrims, and earthquakes had laid the buildings low.

Legend of Pelops

When asked how the Olympic and other games had started, the Greeks, as usual, had a story which explained their origin.

King Oenomaus of Pisa in Elis had a daughter called Hippodamia. To make sure she would marry a man of the right heroic kind, the king decreed that anyone might have her as a bride if he could carry her off in his chariot. The condition was that Oenomaus would chase after the suitor and if he caught up with him would spear him in the back. At the time of the story, there had been more than a dozen unsuccessful hopefuls.

It was very unlikely that anyone at all would win, as the king's horses were magic ones, a present from a god. However, a young man named Pelops, who wished to marry Hippodamia, also had a team of magic horses. At last there would be an even contest.

Perhaps there might have been if Pelops had trusted to luck. He didn't leave things to chance, however. He bribed the king's charioteer to replace the lynch pins on the royal chariot with similar looking ones made of wax.

The chase was to have been from Elis to Corinth, fifty miles away but only a fraction of that distance had been covered when the wax pins sheared through and the wheels came off. The old king died in the accident and Pelops married his bride.

They lived happily for many years. When Hippodamia died, Pelops ordered games and races to be held in her honour as part of the funeral rites.

This, the Greeks were sure, was the beginning of the Olympic games.

Section 7 *Athletics and games*

Olympia

Olympia now

Warrior and horse

Those who know only a little about Greek history often confuse Olympia with Mount Olympus. The last named was the fabled home of the chief gods of the Greek world, whilst Olympia was the site of the games which were held every four years, a period known as an 'Olympiad'. The two places are almost a three hundred mile journey apart. Mount Olympus is in the north east of the country and Olympia is in the south west.

Olympia was in ancient times a sanctuary, or holy place, not a city. It lay in a valley at the foot of Mount Kronos. It was the site of the nationwide 'games' for over a thousand years until the area was abandoned and forgotten.

The river flooded, landslides occurred and both the buildings and sporting arenas became covered many yards deep in sand, silt and mud. So thoroughly were they buried, that there was some argument in the last century as to the actual situation of Olympia.

The first main excavation was done by a German team in 1876. The diggers were partly helped and partly hindered by the habit of local people of raiding the area for building stone. Where the villagers had used spades, the Germans could see what was worth following up. Unfortunately, much of the masonry that could have provided information had gone.

In spite of this, the list of ancient Greek things that were recovered makes interesting reading. To begin with, there were more than 13,000 small or medium-sized bronze objects, many of them given to the local temple as a thanks offering for a victory. There were also over a hundred statues or sculptures, over a thousand terra-cotta objects, together with countless inscriptions and monuments plus an almost unbelievable six thousand coins.

The diggers found the site of the altar to Zeus where athletes promised to keep the rules. They also found the foundations of a huge building, called the Leonidaion, where important guests stayed. There were so many ordinary visitors when the festival was on that they probably had to put up tents or sleep in the open air. Even the officials had tents to live in which could easily be picked out, since they were always made of snow white material. Priests had a permanent building and there was a row of small store houses where gifts offered to the gods were kept.

The stadium, 606 feet long, was where the foot races were run and next to it was the palaestra, or wrestling arena. Somewhere nearby, there must be a horse and chariot racing course. The Germans never found it, in spite of the measurements given by ancient writers – two thousand feet long and several hundred feet wide.

Bronze ram's head

How the Zeus statue might have looked.

The temple of Zeus was represented by foundations and the stumps of fluted columns. At one time there had been a forty-feet high statue of the god made by Pheidias, the artist responsible for the Athena statue in the Parthenon. The Zeus statue was one of the seven wonders of the ancient world.

The excavators even found the workshops where the sculptor had produced his masterpiece. Pieces of the mould for the god's gold draperies were found but more remarkable still was a fragment of broken pottery with 'I belong to Pheidias' scratched on it.

Zeus was not the only god with a shrine and many of the visitors spent as much of their time on religious matters as they did watching the games. As well as priests, worshippers, athletes, trainers, judges and officials, there were horse traders, food hawkers, pedlars of wine with their bulging goatskins plus sellers of trinkets, souvenirs, amulets and small objects to be presented to the various temples.

At one time there had been an oracle at Olympia, similar to the one at Delphi. Legend has it that the games had been abandoned several centuries before Christ and in a period of chaos, civil war and plague, someone asked the oracle what could be done to restore peace. The answer was given that temples and shrines should be repaired and the games begun again.

When they did, a record was kept of the events and their winners. In 776 B.C., when the thousand year cycle of athletic events was started, a man named Coroebus had the honour of crossing the finishing line ahead of his competitors and rivals – the very first Olympic champion.

The Olympic games

Crowning the victors

We've seen that the Olympic games were first recorded in 776 B.C. They took place every four years until 393 A.D. when they were banned. They were therefore held no fewer than 292 times.

At first, this purely religious festival had very little to do with athletics – for several Olympiads, one 200 yard sprint was the only race held. Even when other contests were added, they were packed into one day. Only later were there four days of events with a fifth day for prize giving. The prizes, by the way, were officially nothing more than head bands of wild olive leaves.

However, when the victors returned home, they often found that their own city would give them a pension, free them from paying taxes or feed them for life. Only in Sparta was the reward a place in the front line of soldiers in the next war!

Spectators and competitors were locals at first but eventually they came from all over the Greek world – not just the mainland but also from the colonies along the shores of the Mediterranean. A truce was proclaimed for the period of the games: civil wars stopped long enough for athletes and audience to get to the games and back to their own cities again.

The spectators, perhaps 20,000 in number, tended to group themselves into factions around the running track, rather like the separation of home and away team fans at a modern football match. The crowd was entirely male: women were forbidden to watch or take part. In some places, including Olympia, women held their own games after the men had left.

The opening event was usually a chariot or horse race. Small chariots drawn by four horses raced round the hippodrome (horse track) in clouds of dust. Fitting out a chariot and team was expensive and only rich men could afford to enter. Sometimes as many as forty chariots were ranged along the starting line.

The race consisted of several circuits of the track. A twelve lap race was just over nine miles but few vehicles finished the course. Chariot drivers were the only athletes wearing clothes – perhaps to protect them in case of an accident.

The rest performed naked. Trumpet calls signalled the appearance of contestants, judges and officials through a narrow corridor into the stadium.

The athletes had been practising for almost a year under their own personal trainers. Even the judges were compelled to come early to be taught how to do their job. They had two weapons – a long cane or whip to punish minor law breakers and the power to fine those who cheated deliberately. The money went towards the cost of statues of the gods, on the bases of which were inscribed the names of the cheaters.

On the day itself, the contestants had risen at dawn, prayed to their own gods and promised at the altar of Zeus to keep the rules. Now here they were in the arena. They stripped and oiled their bodies. A herald called for them to take their marks and they fitted their bare toes into a grooved stone which served as a starting block. Another trumpet signalled the 'off' and the athletes ran for all they were worth.

The short race of 200 yards was added to as the years went by. A middle distance race of 400 yards was run and then one of nearly three miles. Curiously enough, there was never a marathon event.

Other events were tacked on – boxing, wrestling, discus and javelin throwing, weight putting and long jumping were some of them. The best performers usually entered the 'pentathalon' (five contests) and the really tough ones the 'pancratium' (all strengths) which was a cross between boxing and wrestling, with almost any kind of attack allowed, short of eye gouging, biting or finger breaking.

If he still had his fingers and could use them, a losing fighter could raise one as a sign of surrender. Sometimes the holds were so fiercely applied that the wrestlers could not free themsleves and had to be prised apart.

When the main events were over there were 'mini' Olympics for the heralds and trumpeters, followed by wrestling and races for boys. One vase painting shows a young hopeful performing the long jump. He is holding 'halteres', or dumb-bell shaped weights in his hands. He held them behind him and threw his hands foward as he jumped. The weights were supposed to increase the length of his jump.

This technique was widely used but not, oddly enough, for the high jump, which was not an ancient Olympic event. Neither, it seems, was swimming, even though the remains of a modern sized swimming pool have been found.

Other games and pastimes

Olympia was not, of course, the only place where an athletic festival was held. There were many of them. In Athens, there were seventy public holidays a year. Every four years there was a week-long festival in honour of Athena at which, in addition to the usual running and jumping, there was a torchlight relay and a boat race.

The three most famous games after the Olympics were the Pythian, or Delphic, the Isthmian and the Nemean. Prizes for winners were laurel leaf wreaths at the Pythian games, pine needle or wild celery crowns at the Isthmian and headdresses of wild parsley or celery at the Nemean.

The Pythian games were held, like those at Olympia, every four years, whereas the festivals at the Isthmus and Nemea took place every two years. Prizes at some of the lesser known festivals included money, jars of wine or oil, clothing and cloaks, or shields, swords and armour.

The same kinds of events were staged at most of the festivals but at Delphi there were also contests to find the best flute or lyre player, or the composer of the best tune. Soon, other organisers were including music in their programmes.

As well as these official contests there were many other sports in which the Greeks indulged which were never, as far as we know, part of a festival's games. We don't know much about them apart from sculptures or vase paintings but it seems they played a kind of football and there is one illustration of a game which looks suspiciously like the start of a game of hockey.

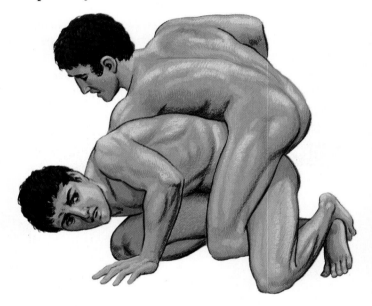

Game resembling hockey

Bowling a hoop

Wrestlers

There were also 'impromptu' games, sports and pastimes indulged in by children and young people generally. It's not surprising to find that many of them are well-known to modern people, as they turn up almost anywhere and in almost any period of history.

Arm wrestling, leap frog, tops, knucklebones (five-stones or jacks), skipping, tug of war, hoop bowling, 'he', 'it' 'touch' or 'tag', pickaback fighting, blind man's buff, marbles and various ball games are just some of them.

A great many of these were played by young children and some, at least, have served to introduce the youngster to the adult sport. A lot of boys must have played with toy bows and arrows, or tried hurling a straight stick as a prelude to javelin throwing. Perhaps they even twisted a strip of thin leather or cloth round the shaft, as grown-ups did, to make their stick revolve rapidly and keep it on course. It's even more certain that boys practised wrestling, just as they've done throughout the ages.

Among young men, there was a game that doesn't seem to have led to anything, except drunken amusement. It was called 'kottabos' and was played at the end of a party.

The performer took a little wine in his wide rimmed drinking cup, and flicked it at a target – perhaps a statuette, or aimed it into an empty bowl. Greeks half believed that if you were good at it, you would be lucky in love.

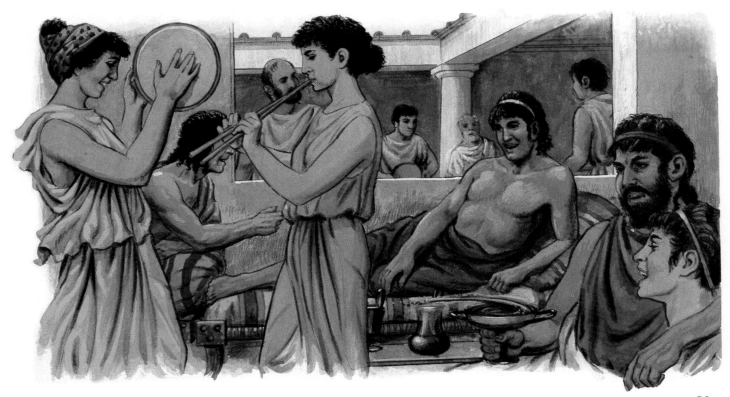

Sparta

A visitor to Sparta was once invited to a meal. He later told a friend that the food was so awful, he could scarcely eat it. When he described the haggis and black broth he was given, the friend said, 'You were lucky to get special treatment – normal Spartan food is much more unpleasant!'

The reason for the poor food was that Sparta was on a war footing and had been for two or three centuries.

In the early days most Greek cities were very similar to each other but after the Dorian invasion, Sparta became the chief city of the newcomers while Athens was the virtual leader of the original Achaeans.

After a while, the harsh conditions in Sparta were relaxed as the Spartans subdued the peoples round about them. For a time cultural activities flourished. Fine pottery was produced and exported; festivals of music and poetry were held.

Then one of the subject peoples rebelled and had to be put down by force. Spartans were afraid that they might not be so lucky next time – unless they were well prepared. After all, the helots, as these semi-slaves were called, outnumbered Spartan citizens by at least seven to one.

The Spartans thereupon made up their minds to turn their city into an army camp. They stopped trading with the rest of Greece and refused to use the new coin money, preferring to go on using iron bars as currency. In future they would only be soldiers and the helots would have to do the farm and other general work. Helots were beaten if the amount of food they produced fell short of what was demanded and could be put to death if they complained.

The only way to maintain this state of affairs was to give every Dorian Spartan a military training.

This began more or less at birth. If a boy baby was weaker than average anywhere in Greece, his father had the right to refuse to bring him up and the wretched child could then be abandoned to die of exposure or be killed by wild animals. In Sparta, the father had no choice – the decision was made by a board of officials and there was no appeal.

Even if the boy was accepted, he was taken from his mother at the age of seven and sent to a military school.

Spartan soldier

There he was subjected to an extreme form of discipline. He ate very poor food, slept on heaps of rushes, wore a thin tunic winter and summer and went barefoot even if there was snow on the ground.

Boys often went hungry and had to steal food. This was encouraged as good practice for the soldier's trick of 'living off the land'. However, if a food thief was caught he could only expect severe punishment and the jeers of his friends.

Spartan boy stealing food

Spartan boys must never show emotion or any sign of weakness. There were even whipping contests to see which youngster could stand the most punishment without crying out. There were cases of lads being beaten to death sooner than show that they were hurt.

The endless military drills and exercises were punctuated every week or so, when the boys were paraded and inspected to see if there was any spare fat on them. The training had no let ups and there were no holidays.

When the youths grew up and began to think of getting married, they would remember that no one was allowed to marry until he was twenty, nor could he live with his wife for another ten years after that.

It is interesting to compare this sort of career with that followed by a young man in Athens (and other Greek cities). The Athenian might or might not have an ordinary education to the age of 18. Thereafter he did only two years military training, compared with 23 years in Sparta.

Sparta itself had no walls because it was thought that the citizen would fight all the harder when he was attacked if he knew there was no other defence. For a similar military reason, no lights were to be shown in the city at night, so that people could get used to moving or even marching in pitch darkness.

Little evidence of a cultured Spartan civilisation can be found from this rigidly controlled age – no great art in the form of sculpture, architecture, poetry or drama. Athens showed the world the first workings of democracy by way of freedom of speech and thought plus the ability of each Athenian citizen to express his own personality in his own way.

Athens and Sparta were rivals for the leadership of the Greek world – first one was on top and then the other. Eventually, as we shall see, the rivalry became so intense that a civil war broke out.

Boy being checked for excess fat

Why the war started

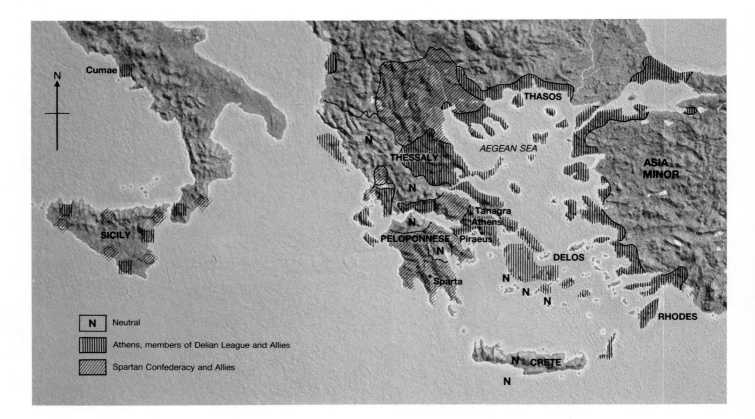

It was probably the defeat of Persia that led to the Peloponnesian wars. The Peloponnese was the southern part of Greece, controlled by Sparta. It was named from the legendary king Pelops (see p.76).

Athens to the north had made certain that Greece was not occupied by the Persians and wanted to ensure that they never did so in future. So they began to recruit allies, each of which promised to help any of the others if they were attacked.

This joint alliance was called the Delian League because it was originally based on the island of Delos, lying half-way between Greece and Asia Minor.

The trouble was that not every colony or city state wanted to join the League. Those that were reluctant were forced in one way or another to become 'allies'. The unwilling members quickly sank to the position of underdogs, to be ordered about rather than consulted as friends or equals.

By the time Athens under Pericles had gathered together most of the mainland city states, the Aegean islands and the colonies in Asia Minor, the whole thing was much more like an empire than a partnership. Severe measures were taken against any city that tried to break away. Athenian leaders chose to look upon these as traitors who were about to join the enemy, even if they were merely fed up with being told what to do by Athens.

One such was Thasos, an island off the northeast coast of Greece. In 463 B.C. Athens sent war galleys to teach her a lesson. Thasos' fleet was defeated, her ships confiscated and her defences torn down. She did what others had done when faced with the same problem – she asked Sparta for help. Unluckily for Thasos, Sparta was too busy to do anything – all her attention was fixed on a revolt of her own slaves.

Incidents like these angered the Spartans. Nor were they the only ones to be annoyed. Many of Athens' subject cities complained that the tax or tribute money which was supposed to be spent on ships to protect them from attack was actually paying for city improvements in Athens itself. A leading Athenian might have replied that the complainers had neither been occupied or raided, so what was wrong?

It wasn't long before Athens seemed to be looking to extend its empire into the western, as well as the eastern, Mediterranean. This, too, made Sparta uneasy, for their own corn supplies came from Sicily where there were many Greek colonies, as there were in southern Italy and the Mediterranean coasts of Spain and France – Marseilles started as a Greek settlement.

Sparta found that she had to build up her own league of friendly cities. In 457 B.C came one of the first actual clashes between the two rivals. A little to the north of Athens, the armies of the two 'super-powers' of those days met at a place called Tanagra. It should have been an easy win for Sparta as their army was far stronger than that of Athens, which was basically a naval power. However, the Spartans only just won in spite of the cavalry from Thessaly changing to their side in the middle of the battle.

This was an unusual event: fighting between the two cities had been rare; now it was to become more common. Sparta's citizens grew convinced that an all-out war was inevitable.

Athens was quite content to sit behind her defences – the city was surrounded by walls, her port, the Piraeus, also had walls and the two places were connected by a road with a high wall on each side. The Athenian fleet was large and strong. Food supplies were protected, so what could happen to Athens? Sparta's army was the strongest in Greece but surely, if it tried to attack Athens, it would soon exhaust itself in the countryside, wouldn't it?

Sparta declared war and Athens ordered all the countrymen to leave their villages, bring their valuables and shelter inside the walls.

Athens and its harbour, Piraeus, showing defensive walls

Soldiers and battles

Our knowledge of the weapons soldiers used and the armour they wore comes from two different sources. On the one hand we have brief written descriptions and representations of fighting men in statues and painted pottery. Our second source is an unexpected place. We have to journey to Olympia, the site of the ancient games.

The arenas were only occupied intensively for a week or two every four years. Those sent to get things ready found that the wells dug last time for drinking water had mostly caved in. It wasn't worth the effort and expense of providing wooden shuttering or a brick lining for something so rarely used, so the advance party cleared the ground by sweeping everything into the old wells and filling them in with the sand dug out of the new ones.

Modern excavators found over 150 wells and the 'rubbish' they contained included old war trophies which had hung on poles until the wood rotted. That was when the orderlies tipped them into the holes and buried them.

Warrior departing for battle
Shield boss

Corinthian helmet
Captured Persian helmet

Phalanx

The war trophies were mainly helmets but there were other military bits and pieces – shields, breastplates, shinguards or greaves and protecting plates for foot and arm. One helmet is inscribed, 'A helmet of the Medes, taken by Athenians' and may be a relic of the battle of Marathon. Another one bears the name 'Miltiades', the actual commander at the same battle.

Back in the days of the Trojan war, the fighting was largely a matter of single combats between warrior heroes who rode to war in chariots. At the time of Pericles, campaigning was usually in the summer months and a whole war might depend on the outcome of a single battle.

The Mycenaeans seem to have measured the riches of a man in weapons and it was not unusual to find a burial accompanied by twenty or so bronze swords. Certainly, by the time of the war against Sparta, wealth determined what sort of soldier a man was. A land owner would be expected to provide a horse, in addition to the arms required of one not so well off.

Even though the bronze age had long given way to the iron age, many weapons and body protecting plates were still made of bronze. There are examples of bronze swords with iron edges. Iron was heavier than bronze and if made (perhaps accidentally) with the right proportion of carbon to iron, the resulting steel would take a better edge.

All sorts of weapons were employed – swords, spears, bows and arrows, daggers and even lariats which some members of the Persian army used. In both Athens and Sparta, the vastly lengthened lance (perhaps 15–20 feet long) was the main weapon of the phalanx.

This was a square formation of foot soldiers, or hoplites. The first four or five ranks could point these lances forward, presenting a moving fence of spikes. Phalanxes might be strung out in various formations across a road, thus blocking it, or plugging a narrow valley.

There was little need for complicated plans of transport or supply: the men brought their own equipment, plus enough food to see them through a few days' campaigning.

Nor was there much scope for clever tactics: the normal plan was to 'steam-roller' your men forward, meeting the enemy head on and hoping that he would be the first to break formation and run.

Cavalry was rarely used in the main battle, but was reserved to cover the army's flanks or to take part in chasing a beaten enemy.

Apart from battles, an army might be ordered to go crop raiding or to besiege a fortified town. The last mentioned was hardly ever successful, in spite of the use of siege engines of assorted types. After all, the Trojan war had lasted ten years and the defenders only lost to a trick in the end.

Athens is conquered

'My name is Patroclus. I lived through this war, so I reckon I'm the best one to tell you about it. It lasted 27 years from 431 to 404, in your method of dating.

'Sparta, as expected, besieged us in Athens during the summers and went back home when the weather got bad. Pericles knew that our army was no match for the highly trained Spartans, but while the navy protected our food supplies, he was content to let our enemy waste his strength in useless campaigns in the nearby countryside.

'Unluckily for us, a severe bout of plague not only killed many of our citizens, it also carried off Pericles in the second year of the war. Those who took his place were not so wise and decided that the best way to beat the Spartans was not just to sit behind walls but to attack their friends.

'In the third year of the war, the island of Lesbos decided to drop out. We were so annoyed that some of us pressed for a fleet to be sent to massacre all the islanders. Eventually it was agreed that we should seize the deserters' fleet, pull down their walls and sell them into slavery.

'After ten years, both sides were exhausted and a peace treaty was signed, promising no more war for at least fifty years. The next thing we knew was that some more allies dropped out with others changing sides. Whether that was the cause or not I don't know but skirmishing broke out and the war began again the following year.

'With our new friends we felt up to tackling the Spartan army but we were over confident and we lost. Our new allies melted away. Then, two years afterwards, the leaders decided, unwisely, as I thought at the time, that we'd send a fleet to attack pro-Spartan cities on the island of Sicily. A man named Alkibiades was to lead it.

'He was young, handsome, intelligent and charming and seemed to be the ideal choice. A pity we didn't know him a bit better though. Just before the ships were to sail, Alkibiades and some of his cronies got drunk and ran riot in the streets, smashing several sacred statues.

'Some of us wanted to condemn him to death for insulting the gods but the final judgment was to take away his command and then to exile him. The next we knew, he'd turned up in Sparta to warn them of our plans.

Alkibiades and his companions destroying the sacred statues

'For us, the attack on Syracuse in Sicily was a disaster from start to finish and we had finally to retire, many of our men being caught and destroyed. Then Sparta set up a base in our countryside not far to the north of Athens, with the result that most of our slaves deserted and ran there for safety.

'The next blow was that we had to close down the silver mines because we couldn't protect them. For coins, the Assembly had to raid temple treasuries. Even after that, we were using copper coins with a thin skim of silver or gold on them.

'Twenty years into the war, a group of four hundred men took over the government and abolished democracy. Mind you, they only lasted three months before we threw them out. Then, would you believe it? Sparta asked for peace and we turned them down!

'By now the war had declined into a string of naval engagements. Our merchant fleet was due in from the Black Sea loaded with corn – very important to people like me who had been living on free hand-outs of food nearly all our adult lives.

'We heard that Lysander, the Spartan commander, had left to try and catch the food ships. We weren't too worried and sent our navy to engage him. After it had left, some of us felt a bit uneasy when we discovered that there were only half a dozen ships left in the harbour.

'Well, of course, we lost the war through sheer carelessness. The crews of our fighting ships were ashore foraging for supplies when Lysander and his Spartans caught them. There wasn't what you could call a fight – it was simply murder. You can guess the effect of this dreadful news in Athens.

'We knew that without grain from the Black Sea, we'd have to give in. We sent men to ask what terms the Spartans were offering. It was months before the answer came and by that time thousands had died of starvation. No food came in: Lysander had anchored 150 ships just off the Piraeus.

'Sparta's conditions were known at last and pretty terrible they were. We had to agree to being ruled by our enemies, to hand over all but twelve of our ships and to tear down all our defensive walls. We had no choice and did what they told us.

'The war was lost for Athens but scarcely won for Sparta. Nearly thirty years of conflict had left the whole of Greece bleeding. Although both we and Sparta recovered a little in the next few years, neither was ever again to attain the power and fame it had once known.'

The defeated Athenians demolishing their defensive walls.

91

Section 9 *Alexander the Great*

Early days

Alexander was born in the year 356 B.C. at Pella, the capital of Macedonia, the son of Philip, king of Macedonia and of Olympias, his wife.

From his father, Alexander inherited bravery, an athletic body, plus a good deal of intelligence and common sense. From his mother he got his extraordinary good looks, a romantic frame of mind and an acceptance of superstitious beliefs. Olympias may have been a priestess of a nature god religion. One of her fellow priestesses once told Alexander that he would never be a loser.

At the time of his birth, Alexander's father, Philip, was busy turning his small, semi-barbaric kingdom to the north of Athens, into the ruling power in Greece. After the ruinous war between Athens and Sparta, Thebes had become the leading city.

Philip slowly pushed out his boundaries, picking off one city state after another. His possible victims quarrelled with one another, each thinking that Philip would never attack, that he would be satisfied with what he had and they made no preparations against attack until it was too late.

Philip, unlike his son, cannot have been a very pleasing sight – he had a crooked back, a limp and only one eye – all the results of war wounds. It must be said, however, that he wasn't a complete barbarian. To start with, Macedonians spoke Greek, even if with a strong accent, and the king himself so admired the culture of Athens, that he appointed Aristotle, the great philosopher, to be tutor to his son and heir. His influence gave the boy a love of Greek literature and Alexander is said to have carried a copy of Homer's *Iliad* on all his future campaigns.

A story is told of how the twelve year old Alexander acquired his famous horse, Bucephalus, which carried him 'to the ends of the earth'. A horse dealer arrived one day with animals for sale. Philip was still a soldier and a countryman at heart, so he attended the sale himself and took his son with him.

One of the beasts didn't seem to have been broken. It had wild, rolling eyes and shied several times. Philip at first would not consider buying it until Alexander begged the king to allow him to try and ride the horse. Alexander had noticed that the animal seemed frightened of its own shadow, so he turned it towards the sun, all the while murmering calming noises and patting its neck.

Soon he managed to get on its back and before long, he had it tamed. From the shape of its head, Alexander called it 'Bucephalus' – from two Greek words, 'bous', an ox, and 'kephale', a head. Philip bought it cheap from the dealer.

When it eventually died at the great age of thirty, Alexander had reached the Punjab in what is now Pakistan and he built a city called Bucephala in its honour.

In his youth, Alexander rarely drank wine and seldom overate. He was therefore, even at the age of sixteen, in superb condition to take over command of the army while his father was away. In double quick time the young man had put down a rebellion of hill tribesmen on Macedonia's northern frontier.

When Philip returned to fight a battle against the combined Greek states, he took over the supreme command again but Alexander was put in charge of the left wing of the army. Philip led the right wing. Against Alexander was the famous 'Sacred Band', the elite of the Theban army. The young Macedonian led the charge which smashed them to pieces.

On the other side, Philip pretended to retreat. The Athenian soldiers ran after his men who then turned and caught them in a trap. This battle took place at Chaeronea on the plains of Boeotia and marked the end of Greek resistance to the king of Macedonia.

When the fighting was over, Philip embraced his son and told him that he'd have to conquer foreign lands, for Philip's new Greek kingdom, large though it was, would not be big enough for Alexander.

Philip was murdered at a relative's funeral and Alexander succeeded him as king. He was just twenty years old.

A corselet and shield found in the grave of Philip

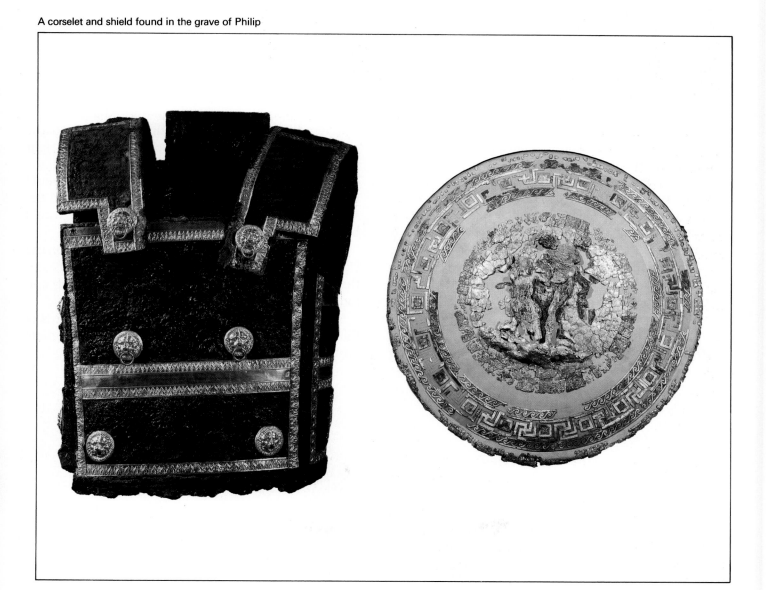

The army

Alexander did not have to reorganise the Macedonian army – that had already been done by his father. Philip as a young man had seen what a poor state the knowledge of military matters was in and when he became king he had determined to do something about his own army.

We saw on p.89 what the normal practice was – each city state made use of the phalanx as a battle formation. This was a group of men arranged in a block and armed with long spears. The most successful were those whose soldiers were the more determined, strong and courageous.

The commonest type of warrior was the hoplite, an infantryman. Cavalry was almost unknown before the Persian wars, except in northern Greece. The hoplite usually carried a round shield, a spear and a sword. These, together with his body armour, weighed over seventy pounds. Thus, the phalanx wasn't exactly designed for speedy attacks.

There were points to be made both for and against the phalanx. In its favour was the fact that magnificent discipline and loyalty kept it fighting. Against it was its weakness when tackled from either flank, particularly the right, or shieldless side.

So Philip made his Macedonian phalanx a defensive block rather than an attacking one. He gave the men even longer spears, up to twenty feet, in fact, and added cavalry wings.

To fight a battle, he preferred to have six divisions of 5000 men each in the centre plus blocks of men with an attacking role, armed with spear and sword, heavy horsemen on either side and fast, lightly armoured cavalry beyond them again. There were also groups of bowmen.

Philip had learned from the Thebans that it was better to pick a weak target rather than use the Spartan method of advancing against whatever was in the way. The Thebans, however much use they made of mounted soldiers, usually chose the phalanx to destroy the target. This was not Philip's solution. He preferred to send in whichever of his groups stood the best chance of winning, having taken into consideration the nature of the ground and the strength and position of the enemy.

This then, was the army that Alexander took over when his father died. He added another refinement – the attack from the rear. It was managed like this.

The advance was not in a straight line parallel to the enemy forces but on a slant or 'echelon', with the right wing leading. That wing would make contact with the opposing forces first. The rest of the enemy would naturally continue to advance until they reached the Macedonian left wing. There the light cavalry would hold them by charging repeatedly.

At the same time, Alexander's heavy cavalry on the right made a tremendous effort and smashed through the lines in front of them. They then wheeled left and took the foe from behind.

These tactics were never faulted and the words of the priestess that Alexander would always be victorious proved true.

Macedonian phalanx

Alexander and his army fording a river.

Battles and sieges

From the previous page, it would seem that Alexander relied on pitched battles for his conquest of the known world. This is not really true. Of course, he certainly did pit his Macedonian troops against those of rulers who stood in his way but he relied just as much on siegecraft.

Before his father's time, the Greeks had known little of the art of capturing a strongly fortified city. Philip realised that without reliable siege methods, Greece would never become a nation, for the separate city states might have to be beaten one after the other. How do you do that if you can't get a walled city to surrender?

Philip knew full well that the inhabitants of a city threatened with a siege only had to retire behind their thick stone or brick walls and wait for the enemy to go away. Only starvation could make them give in and few attackers had the time or patience for that kind of operation.

He therefore built up a corps of siege engineers and brought in ideas of wall attacks from farther east, where they had been known to the Assyrians and other ancient peoples. Philip was never very successful with his ideas, in spite of the fact that he was probably the first European ruler to introduce missile hurlers such as mobile catapults and ballistae.

Alexander used these machines and also battering rams, siege towers and pontoon bridges to cross supposedly 'uncrossable' rivers. Siege towers on ships and an artificial pathway, or mole, were features of Alexander's seven month siege of the city of Tyre. Ever scornful of danger, the young king led the final assault in person.

To conquer Persia, which was his life's ambition, Alexander had to defeat 'the Great King' – Darius. The Persian ruler had not thought it worthwhile to confront Alexander and took no part in the first battle fought by the two old enemies on Persian soil. The Macedonians won handsomely and sent huge quantities of booty back to Greece.

Before the next important battle, Alexander was shown the famous Gordian knot. A chariot's pole was tied to a post in an ancient temple. He was told that

A siege tower with catapult

A battering ram

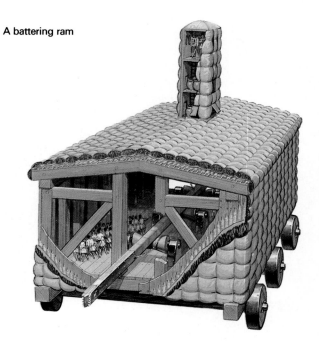

whoever could undo the knot in the leather would rule Asia. Alexander solved the problem by slicing through it with his sword.

The battle of Issus started with Alexander's troops strung out according to his pet plan, with the right wing advanced. One writer says there were over half a million Persians ranged against him but this must be a wild exaggeration.

Alexander led his men forward at a run, trying to avoid the enemy's arrows. They got to within striking distance of the Persian king's chariot. After many of his defenders had fallen, Darius fled the battlefield. Only a little while later, his men followed him and ran away.

Before the final battle with Darius, Alexander swept down the coast, into north east Africa and freed Egypt from Persian rule. Then he struck eastward, crossing the rivers Euphrates and Tigris with 40,000 infantry and a large number of mounted men.

They met the main Persian army at a place called Gaugamela, now in Iraq. The night before, the Macedonians had been astonished at the size of the enemy army, revealed by twinkling camp fires in the dark. The morning showed that it had not been an illusion – the enemy numbers were enormous, with soldiers from every part of the Persian empire, plus war elephants and even chariots with sharp scythe blades on the wheel hubs.

In spite of all this, Alexander threw his cavalry into a gap that had appeared in the opposing lines. They made straight for the unmistakable figure of Darius in his huge, decorated chariot. Hand-to-hand fighting raged towards the Persian ruler who, suddenly afraid, turned and galloped off.

Some time later when Alexander's men had swept away the tattered remnants of the once proud Persian army, the pursuing Macedonians caught up with Darius's chariot. The king's dead body was lying half out of it, apparently murdered by his own men.

Now, for the first time, Alexander could consider himself the ruler of the Persian empire.

Persian war elephant

The murdered Darius

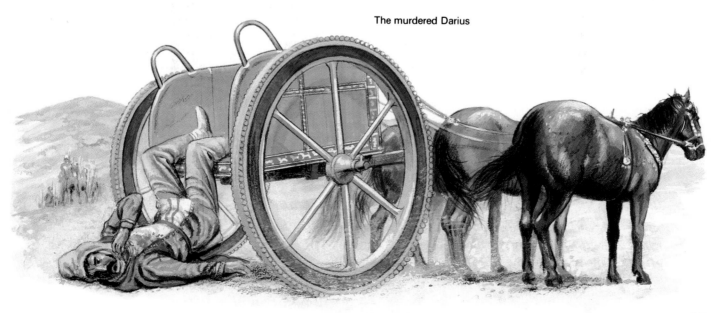

Persepolis

'So this is what it feels like to be an emperor!'

Even before Darius had met his death at the hands of his own officers, Alexander had entered the magnificent Persian capital that they called Persepolis. In Greek, the word means no more than 'City of the Persians'.

Alexander sat on the throne of the Persian kings and is reported to have said, 'So this is what it feels like to be an emperor.'

Already he was secretly making plans to unite Persia and Greece under one king and perhaps to extend the boundaries even farther to the east and south.

At the same time, Alexander's relationships with his old comrades were beginning to change. Once he had been a soldier among soldiers, enjoying the fighting, taking a leading part in the battles and spending the nights yarning and drinking with his friends. Now he was gradually realising that a king does not do such things.

The spell of the east was increasing its influence on him. Eastern luxury was soon to be considered normal – a strange idea for a man from a rough, hard and poor country.

However, the business in hand for the Macedonians was the systematic looting of the treasuries of Persepolis. All kinds of valuables in gold and silver – cups and plates, jewellery, statuettes, crowns, tiaras and many other things were packed into containers.

These were then slung on the backs of mules and camels, lashed into place and then started on their long journey back to Greece. Among the objects sent, and ones that gave particular pleasure, were Greek statues that had been looted and captured during the Persian invasion of Greece – probably taken when the Persians had sacked Athens.

The spoils of Persepolis being taken to Greece

It is said that the pack animal journeys were countless and this statement, for a change, was probably not much of an exaggeration. In addition to the things taken away, there are still in existence the remains of luxury objects which the Greeks smashed.

It was also said that Alexander and his men spent the night in drunken revels and that as a result, a fire was started accidentally which swept through the buildings of the Persian capital and utterly destroyed it.

The fire may have been accidental but it is just as likely to have been set deliberately. There are two reasons given for this. The first is that it was a signal to the Persians and their subject peoples that an age had come to an end.

The second, and probably stronger reason, is that this was an act of revenge – a reprisal for the Persian destruction of Athens and other parts of Greece.

Whatever the truth of the matter, no one will ever know now which it was – an accident or revenge. The fact is that Athens is still a thriving, busy city and Persepolis is nothing but a few forlorn ruins.

Persian gold armlet

Persepolis on fire

Alexander's last battle

Alexander led his army on forced marches, fought few pitched battles but did a good deal of skirmishing and put down many minor and major rebellions. Farther and farther the army went – right across the western part of Asia and down into the Indian sub-continent.

Constant campaigning, more and more bouts of violent drinking and lack of rest began to undermine Alexander's health. In spite of this he pressed on with what he thought were necessary reforms. In order to get obedience from his conquered subjects, he came to the conclusion that he must be considered more than merely a king but a living god instead.

He began to wear Persian clothing and behave the way he thought a god would conduct himself. During one of his frequent drinking bouts, one of his old comrades made a joke about Alexander's 'pretensions' to be divine. Alexander took offence and ordered him to bow down in an attitude of worship. The man, who had once saved Alexander's life, laughed and refused. The king scarcely paused: he caught up a spear and ran his friend through, killing him instantly.

Things like this now seemed to be of little importance compared with his ideas of world conquest. He realised that the only part of the old Persian empire he had yet to win was India.

So, in the early part of the year 327 B.C. he set out with an army now largely composed of local soldiers, although some of the Macedonians who had been with him from the beginning also went with him. He paused before crossing the high mountains guarding the approaches to India and subdued the country round about. During one of these campaigns, he captured a princess named Roxana and made her his wife.

Then he led his men along the Kabul valley and through the Khyber Pass, fighting fierce tribesmen all the while. They moved on through the Punjab and came to the River Indus.

Putting down local rebellions, making friendships and treaties and hoping perhaps to found more cities, many of them named after himself (there were already over seventy of these stretching all the way back to Egypt), he intended to press on eastward.

In fact, he needed to reach the river Hydaspes, a hundred miles away before the rains came and the river became uncrossable. Beyond the river he was sure, was the fabled 'Ocean', a mythical world river which many Greeks believed to encircle the entire earth.

When the army arrived on the banks of the Hydaspes, it was already beginning to rise. Alexander sent for vessels to be brought in pieces on wagons from the Indus. Then he confused the enemy king, Porus, by dividing his army into separate detachments and ordering them in different directions, so that the enemy didn't know what to guard against.

The boats were put together some miles from Porus's main camp. Alexander's army made their way

secretly at night, the noise of their passing being drowned by a violent thunderstorm.

Porus marched off to meet him with his chariots and elephants. Alexander put into operation a variation of his wheeling and striking from the rear tactic. On top of this, his infantry hurled javelins at the elephants and his archers fired arrows at them until the huge beasts, maddened by the attacks, charged dangerously against friend and foe alike.

Eventually, Porus, surrounded by his foes, surrendered to prevent a massacre. The battle, last of the four fought by Alexander, had lasted almost eight hours.

Alexander made it known that he intended to continue his progress eastwards. At last the army refused. The Greeks still with him had been away from home over eight years and had marched a distance very nearly equal to half way round the world. Alexander sulked for three days but even he had to give in at the finish.

The orders were given to turn round and head back towards Europe. The Great Expansion was over.

The elephants, maddened by pain, became uncontrollable.

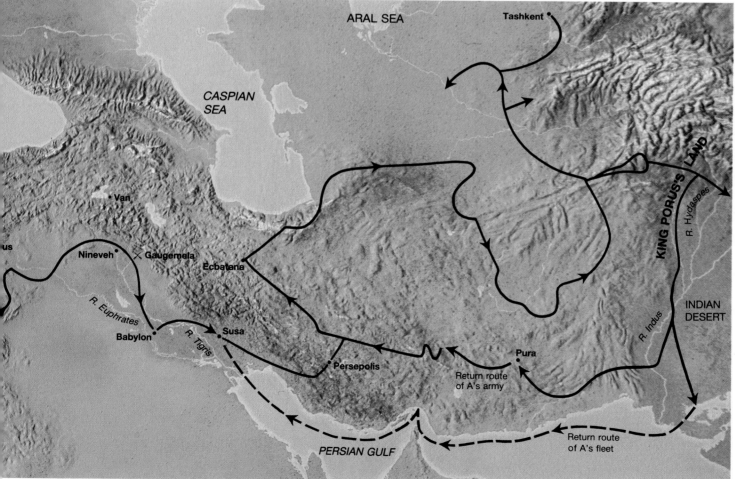

Death of a conqueror

Alexander did not lead his men back the way they had come. He had previously given orders to his shipwrights to launch eight hundred vessels on the Indus and he now proposed to go on board and sail down the great river to its mouth. Admittedly he was going south westward rather than to the east but the great conqueror still wanted to explore places unknown to himself.

As they descended the river, they became involved in many small wars and sieges. At one such siege, Alexander, in a fit of bravado, leapt with two other soldiers from the top of a scaling ladder into a hostile town. There he was hit by an arrow and would have died if one of his companions had not protected his fallen body with a shield until help arrived.

They reached the delta of the Indus and it was decided to explore a new route back home, the fleet sailing up the Persian Gulf and Alexander leading the 15,000 strong army along the shore line.

Difficult conditions forced the marchers to move inland and cross a desert where they almost ran out of food and water. Hostile tribes had to be put down before they could get even as far as Susa.

When Alexander found that some of the men he had appointed to govern during his absence had behaved badly, he had them put to death. On the positive side, more than 10,000 of his Macedonians had married native wives, and 30,000 of the local boys who had been taken and trained in Greece had now

Alexander is wounded

There were frequent exercises to keep the troops in practice.

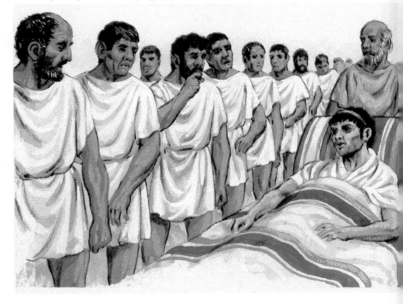

returned as young men and been absorbed into the new army.

Alexander had a dream of a lasting peace between all men – Macedonians, Greeks, Persians and other barbarians – perhaps even extending to those who lived at the western end of the Mediterranean.

In the spring of 323 B.C. he returned to Babylon and began immediately to make plans for further exploration and conquest. His ideas for some sort of world government never really got off the ground.

He never stopped working away at his projects. Even though no expedition was under urgent consideration, Alexander went on organising troop exercises and reviews, sail-pasts and mock battles on the river, giving prizes to those who did the best.

The story goes that he had caught a slight chill and tried to shake it off by heavy drinking. However, the chill turned to a fever and even his magnificent constitution was unable to fight off the infection. Years of constant physical and mental effort had weakened him, as had the numerous wounds he had suffered in various military engagements.

He grew worse and when it became obvious that he wasn't going to recover, his physicians allowed his old Macedonian army companions to file past his camp bed to say their goodbyes. By this time, he was too far gone even to speak and could only greet each man with a faint movement of a hand or a glance from his eyes.

About a week before midsummer day in the year 323 B.C. he died. He was thirty-two years old and had reigned not quite thirteen years. During that time he had conquered a good half of the known world. His body was taken to Egypt and buried in his own city of Alexandria.

103

The empire is divided up

It was inevitable that Alexander's empire would break up, although it didn't happen straight away. Initially his officers had agreed to wait for his child to be born to Roxana, his wife. If it should be a boy, they would make him heir to his father's crown.

The ordinary Macedonian soldiers, on the other hand, preferred to hand over the empire to Alexander's half brother, who, unfortunately, was also a half wit. Eventually an agreement was reached under which both should rule jointly.

They might have saved their breath and their efforts. Not many years later, the son of Antipater, one of Alexander's generals, had both the heirs murdered, together with Olympias, Alexander's mother.

Murder of Olympias

Antipater himself took over the government of the European part of the empire. It was easy for him to do this, for Alexander himself had appointed him to run mainland Greece while the Macedonian army was away conquering the world.

Almost as painless was the siezure of power in Persia and Babylon by General Seleucus. He even extended his territory by adding Syria to his realm. He set up a new capital at Antioch on the River Orontes and his descendants ruled this part of the world for many years.

The third of Alexander's most powerful generals, Ptolemy, began to reign as king in Egypt, over part of the north African coast and also the Holy Land. His descendants included the famous queen, Cleopatra.

The boundaries of these three major areas, not to mention the many smaller ones, varied from time to time as the newly made 'kings' fought each other to enlarge or protect their kingdoms.

The next forty years saw the continuation of these struggles and during this time, the same kind of thing, although on a smaller scale, was going on in Greece itself. Alliances and leagues were formed, split apart, and reformed. Neither Athens nor Sparta ever again became leader of the Greek city states, although Athens in particular continued to be celebrated for her culture and philosophy.

We ought to take a look at just one of the many cities founded by the conqueror to see how Greek thought was flowering after being planted in foreign soil.

Alexandria in Egypt had become one of the leading Greek cities of the ancient world. Here was founded the Museum – almost the equivalent of a university. The city was also home to the best poets of the age, or at least those writing in Greek.

Among the thinkers of Alexandria who made important advances in science, mathematics, geography, astronomy and the arts of healing were Hero, Archimedes and Eratosthenes.

Hero probably lived somewhat later than the other two mentioned. We know from his writings that he had invented an early coin-in-the-slot machine for dispensing wine and also a primitive steam engine.

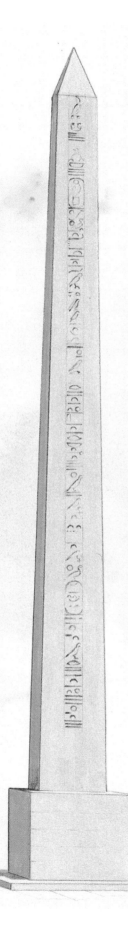

Hero's steam engine

Eratosthenes measuring the shadow of an obelisk

Archimedes, a native of Syracuse in Sicily, studied at Alexandria where he made great discoveries in science generally and physics in particular. He invented the Archimedean screw or helix for raising water and also (somewhat doubtfully) a method of setting fire to the sails of enemy ships by using huge magnifying glasses aimed at concentrating the sun's rays.

Eratosthenes was one of the first men to discover that the earth was round rather than flat. More important, he hit upon a method of measuring its size. He was told that on a certain day, the sun shone straight down a well at Syene, about four hundred miles due south of Alexandria, whilst an obelisk at the latter place cast a slight shadow.

From these scraps of information, he worked out that the world must measure 252,000 Greek stadia all the way round. This comes to almost 29,000 miles, a figure which compares very well with the modern calculation of about 25,000 miles.

So, in spite of the political collapse of Alexander's empire, Greek ideas, language and culture continued to hold sway over a large part of the known world.

The Romans come to Greece

A Roman soldier of 140 B.C.

Another city was becoming powerful in the Mediterranean at the same time that Athens and Sparta were reaching their peak. This was the city of Rome in central Italy. Until about 200 B.C. Rome was fully occupied in fighting a war against a rival city in North Africa called Carthage.

After Carthage and its general, Hannibal, had been dealt with, Rome was able to turn its attention to the eastern Mediterranean and in particular to one of Alexander's successors, Philip V, king of Macedonia. The Greek king had made certain of attracting Roman notice by entering into a treaty of friendship with Hannibal, the enemy of Rome.

On top of this, Philip had been trying to enlarge his kingdom by expanding into the territory of his neighbours. When he seemed to have designs on nations next door to Rome, or at least, only on the other side of the Adriatic, the Romans decided to interfere.

A three year war ensued, ending with the battle of Cynoscephalae. which almost turned into a farce. Both Rome's and Philip's army had each about 26,000 men – the Macedonians divided into two phalanxes of 8,000 men apiece, plus about 7,000 or so light troops and more than 2,000 mounted soldiers. The Romans had roughly the same number of heavy infantry, 16,500, together with 6,000 lightly armed foot soldiers and perhaps 4,000 cavalry. The Roman forces were commanded by Quinctius Flamininus, who could also call upon auxiliary troops and a certain number of war elephants.

Philip moved, first to one place and then another, seeking a suitably level battle ground, Flamininus followed him. Finally, the two armies, unknown to each other, made camp on opposite sides of the same low range of hills. The name 'Cynoscephalae', mentioned above, is Greek for 'dogheads' and refers to the shape of some of the rocky outcrops.

Next day there was a thick fog and parties of foot soldiers set out to look for each other. After wandering uncertainly for some time, a group of Roman light infantrymen met a similar force of Macedonians. The Romans were driven back, reinforced and driven back again. Luckily for Flamininus, his cavalry turned up in time to prevent a massacre.

When the mist had dispersed, Flamininus began to arrange his troops in battle formation, ordering his right wing to stand still, protected by a screen of elephants, while he personally led his left wing forward. Philip meant to stand his ground but was persuaded to advance, one phalanx leading the other. The command to lower lances was given and the Macedonians drove the Roman left wing some way down the hill.

The Roman right wing with its elephants then attacked the attackers – partly from one side. A Roman officer, seeing the success of this manoeuvre, wheeled some of his men around and took the Greeks in the rear.

The phalanx was not designed for this type of fighting and while the Greeks were trying to get into some sort of defensive order, Flamininus regrouped his left wing and attacked from the front.

The Macedonians could find no answer to this two- or three-pronged onslaught. When he saw defeat staring him in the face, Philip left the battlefield with a small mounted escort.

The Romans lost seven hundred men but the field of Cynoscephalae was strewn with 8,000 Greek dead. In addition, 5,000 of them were taken prisoner.

The Romans won because their army was organised in a more flexible way than the enemy – so that it was possible to alter a plan at the last moment in response to a changing situation.

This was merely the first encounter between the rising power of Rome and the declining power of what was left of Alexander's empire. It was not to be the last.

Greece becomes a Roman province

When the Romans withdrew from Greece in 194 B.C., Antiochus III, king of the Seleucid empire, tried to take their place. Unfortunately for him, his attack on Greece was far too slow and clumsy. It gave Rome time to counter-attack.

The form this counter-attack took was rather like the Roman plan for beating Carthage. Rome wouldn't attack Antiochus's invasion force directly, just as she hadn't tried to smash Hannibal's army in Italy. In both cases, it was decided to counter-invade – in other words, to draw off the enemy from his target by a campaign against his home territory.

The leader of the Roman forces was Lucius Scipio, whose brother Publius Scipio Africanus had been the successful general against Hannibal and Carthage. In fact, Publius volunteered to be his brother's second-in-command. He lent his considerable skill to the task of landing the Roman legions in Asia Minor. Unfortunately he fell ill just before the crucial battle was fought and Lucius was on his own.

To add to Lucius's difficulties, his own army of some 30,000 was outnumbered by the enemy's 60,000 or

more infantry, not to mention another 12,000 cavalrymen. However, Antiochus, instead of remaining in his fortified camp at a place called Magnesia, came out to do battle.

He seemed to be winning for a while but the Romans soon built up such a threatening force at an important point that the king took fright and bolted with the scattered units of his once proud army.

Roman peace terms were that Antiochus should take what was left of his army out of Asia Minor and leave it for the Romans to do with it as they wished. The treaty Antiochus had to sign was finalised as the Peace of Apamea in the year 188 B.C.

The scene shifts back to mainland Greece, as Philip V's son and successor, Perseus, sought to re-establish the complete independence of Macedonia. So, in 168 B.C. Rome sent a force under Paullus to deal with him.

He landed with his troops and took control of a mountain pass not far from the Macedonian camp.

Perseus had to retreat and the Romans caught up with him at a place named Pydna.

Paullus had never seen a phalanx in action before and privately admitted that the glitter of sunlight on the countless lance points filled him with dread. However, nothing of this fear was allowed to show as he gave out his battle orders. Roman troops must advance and push their way into any gaps that could be seen in the phalanx.

The leading cohort of Paullus's troops was almost wiped out but then the phalanx opposing them reached uneven ground and the gaps began to appear. The Romans, sword in hand, threw themselves into the openings, driving deep wedges into the enemy. The Macedonians could manage no counter move and defeat turned into rout when the disorganised phalanx was attacked by the second Roman legion.

Perseus gave it up as hopeless and rode away with his horsemen, leaving the infantrymen to their fate. 20,000 were killed and more than half that number taken prisoner – to become either slaves or gladiators in some Roman amphitheatre. Perseus himself was later captured and died in a Roman prison.

This was the last known use of the phalanx in battle. Just under 150 years after the death of Alexander, his empire had been mortally wounded. Macedonia was divided into four separate regions paying taxes and tribute to Rome. Twenty years after the battle of Pydna, there was a rebellion in Macedonia. It was put down with great severity and Macedonia became a directly ruled Roman province in 148 B.C.

Two years later, there were similar risings in other parts of Greece. These were crushed even more cruelly. Lucius Mummius was the Roman responsible. Corinth was attacked in 146 B.C. and taken. The city was destroyed and the people killed or made slaves.

The difference between the cultured Greeks and the rough and ready Romans was shown when pack animals were being loaded with priceless art treasures from civilised Corinth. An orderly nearly dropped an exquisite statuette. 'Careful,' said Mummius gruffly, 'if you break that, you'll have to replace it.' He wasn't joking but seriously believed that if a work of art were destroyed, you could always get an identical replacement.

From this time onward, there were no more free Greek city states. Most of the rest of Alexander's old empire fell into Rome's hands, bit by bit. The last piece was Egypt which came under Roman rule when Cleopatra and Marcus Antonius were defeated by Augustus at the battle of Actium in 31 B.C.

The battle of Pydna

The Greek legacy

As we have seen, Greece fell prey to Roman empire building. It was almost two thousand years before she regained her freedom – not in fact until 1830, when Turkey, the last occupier, was driven out.

However, the Greek way of life was so attractive to the peoples Greece conquered or was conquered by, that many of the strands of Greek life can still be seen in the modern world.

The list of subjects with a strong Greek influence on our own world is a very long one. We now know that there are many words from Greece connected with the theatre – the word 'theatre' itself, 'scene', 'orchestra' and 'chorus', and it is surely no accident that modern theatres owe a good deal of their layout to the original plans of the Athenian and other classical Greek theatres.

'Odeon' is another Greek word. It was a kind of theatre which staged music and poetry readings. Both 'music' and 'poetry' are from Greek, as are 'harmony', 'rhythm', 'rhyme', 'epic', 'lyric' and 'elegy'.

We've heard about some ancient Greek poetry in stories of the Trojan war but there was much else written in verse. These verses often acted as patterns to later Roman poets such as Virgil and Horace, who in turn influenced English poetry.

'Athlete' is a Greek word; so are 'discus', 'stadium', 'decathlon', 'marathon' and many others. Naturally, we use these classical words because the Greeks were the first to think of the ideas behind them.

When travellers nearer to our own times visited the classical sites, they brought back tales of beautiful buildings and statues. In Europe and America these were copied extensively. There is hardly a major modern city that does not owe something to Greek architects. In England in particular, it was not just the public buildings that were designed to look like Greek temples – even the humble dwelling house of the seventeen and eighteen hundreds might have a temple triangle or pediment over the front door and a doric column on each side of it. Inside, other doors and even fireplaces clearly showed traces of Greek design.

Many terms in science are from Greek, either directly as 'mathematics', 'geography', 'physics' and 'geometry', or they are coined by modern man to name and describe modern discoveries, ideas and inventions. This group includes words such as 'telephone' (distant

The British Museum

Examples of the influence of classical Greek architecture on more modern English architecture.

voice), 'microscope' (small seeing), 'metropolis' (mother city), 'polytechnic' (many arts) and thousands more. Although it is possible to open a good English dictionary at random and not find several words of Greek origin on the page, it is not terribly likely.

It's surprising at times to find out how advanced some Greek ideas were. You may remember Eratosthenes who reasoned that the world was round and who worked out an approximate size: he also drew a fairly accurate map of the earth as known in those days and guessed that it might be possible to sail to India around the south of Africa, or even by going westward across the Atlantic – 'Provided,' he added, in a remarkable glimpse of the truth, 'that there is no large land mass in the way'!

Herophilus of Chalcedon thought that the blood might circulate round the body in the arteries and that perhaps the nerves conveyed sensations to the brain and sent signals back to the limbs. Also strikingly modern is the insistence of some Greek doctors on diet, exercise and hygiene in addition to vegetable based medicines.

Greek styles in pottery and sculpture have been copied in modern times, as have Athenian and Spartan soldiers' helmets – to be seen on the heads of many European mounted soldiers, and even firemen, in the last century.

Probably Greece's greatest gifts to the modern world are to be found in the realms of philosophy, religion and politics. Many patterns of thought laid down by classical thinkers are still followed today.

Although we no longer worship the Olympian gods, their names and deeds are the common heritage of all educated people in the modern world. It's also worth remembering that it was the Greek language which made possible the spread of Christianity and that the Gospels were actually written in Greek.

Undoubtedly we owe the Greeks a great deal – not least in their discovery that it is possible to run a nation on the principle that everyone should have a voice in the government.

'Democracy', too, is a Greek word.

A section of the sculpture from the walls of the Parthenon treasury

Index